Low Temperature Detectors
for Neutrinos and Dark Matter

Low Temperature Detectors for Neutrinos and Dark Matter

Proceedings of a Workshop, Held at
Ringberg Castle, Tegernsee, May 12–13, 1987

Editors: K. Pretzl,
N. Schmitz, and L. Stodolsky

With 94 Figures

Springer-Verlag Berlin Heidelberg New York
London Paris Tokyo

Dr. Klaus Pretzl
Professor Dr. Norbert Schmitz
Dr. Leo Stodolsky
Max-Planck-Institut für Physik, Föhringer Ring 6,
D-8000 München 40, Fed. Rep. of Germany

ISBN-13:978-3-642-72961-4 e-ISBN-13:978-3-642-72959-1
DOI: 10.1007/978-3-642-72959-1

Softcover reprint of the hardcover 1st edition 1987

2153/3150-543210

Preface

For several years the communities of particle- and astrophysicists have shared a common interest in investigating intriguing topics like

a) the solar neutrino puzzle: What happens to the neutrinos on their way from their origin in the sun to the earth? Do neutrinos have a mass and if so, do they oscillate via a mechanism as described by Mikheyev and Smirnov?

b) the growing evidence that our universe is filled with much more matter than is visible: What is the nature of this dark matter? Is it of baryonic origin or does it contain a substantial non-baryonic component, for example weakly interacting stable massive particles like heavy neutrinos, or supersymmetric scalar particles like sneutrinos, photinos etc.?

c) supernova explosions: What do neutrinos tell us about such explosions and vice versa? On the one hand, neutrinos are messengers from the innermost core of a supernova which probe the thermodynamics of the collapsing matter. On the other hand, supernova explosions may yield information about neutrinos. For example, from the energy and arrival time spectra of supernova neutrinos information on the neutrino mass can be inferred.

All these questions link particle physics and astrophysics. Their investigation requires unconventional detection methods employing new techniques which are presently under development.

It was the purpose of this workshop to discuss new developments in the detection of low energy neutrinos and dark matter. The workshop was held on March 12 and 13, 1987 at Ringberg Castle (Conference Center of the Max-Planck-Society) near lake Tegernsee in the Bavarian Alps and was attended by 43 physicists working in different fields such as solid state, low temperature, astro- and particle physics. It was organized by the Max-Planck-Institut für Physik und Astrophysik.

The first day of the workshop was mainly devoted to techniques using superheated superconducting granules (SSG). For solar neutrino detection two main ideas were discussed. The first, developed by Stodolsky and Drukier, uses coherent neutral current neutrino-nucleus scattering. This method has the advantage that the cross section is about 1000 times larger than the cross sections of other processes like, for example, inverse beta decay. Thus, a SSG detector with a weight of a few kilograms would measure the same event rate as a multiton detector based on other processes. The second advantage is that the SSG detector responds to all neutrino flavors equally. The principal difficulty with this method is, of course, the detection of a very low nuclear recoil energy E_A. (In the limiting case, $E_A \sim 1\,\text{eV}$ assuming Sn grains, for solar neutrinos with $E_\nu = 0.4\,\text{MeV}$.) In a uniform heating model most of this recoil energy will be transferred into heat, leading, due to the low specific heat at very low temperatures, to a measurable temperature change ΔT of the grain. This temperature jump can flip a grain from the superconducting to the normal state. The grain flip can be detected with a pick-up coil which measures the flux changes due to the disappearance of the Meissner effect in a magnetic field. Tiny Sn grains of $2\,\mu\text{m}$ diameter have to be cooled down to $T = 50\,\text{mK}$ to be sensitive to $\sim$400 keV solar neutrinos.

The other solar neutrino detection method using indium granules and inverse beta decay reactions was discussed by G. Waysand. This method has the advantage that the requirement on the sensitivity of the granules is relaxed by several orders of magnitude (1eV in the coherent scattering case versus several 100 keV in the case of inverse beta decay), but a heavy weight of $\sim$4 tons of indium granules, cooled down to 1 k, is needed.

For a study of the solar neutrino puzzle both approaches are complementary, since the first method is neutrino flavor independent (insensitive to neutrino oscillations) while the second depends on flavor (sensitive to neutrino oscillations). SSGs also offer a way to study magnetic monopoles and dark matter, which has raised so much interest in cosmology. SSG might also be used for double beta decay experiments (reported by P. Pacheco).

Two further ideas using superconducting detection materials were discussed by O. Liengme and G. Vesztergombi.

Considerable progress has been made in understanding the basic properties of superheated granule detectors. Results of irradiation experiments with α, β, and γ sources (L. Gonzales-Mestres, D. Perret-Gallix, K. Pretzl, and A. De Bellefon) have shown that the superconductivity of the grain starts to be broken locally around a region where an incoming particle deposits most of its energy. This leads to a fast local break-up of Cooper pairs, which apparently occurs before the grain is globally heated by phonons. Experimental results on granule readout systems like SQUIDs (superconducting quantum interference devices) and linear amplification systems were presented by A. Kotlicki and G. Waysand respectively.

The second day was mainly devoted to other techniques like bolometry, superconducting tunnel junctions, ballistic phonons, and the detection of solar neutrinos in superfluid helium.

B. Sadoulet reviewed the cryogenic detector efforts in various U.S. laboratories. The physics motivations are to detect astrophysical X-rays with 1 eV resolution, dark matter, solar neutrinos, and double beta decay. The detection techniques under study use SSG, ballistic phonons in connection with tunnel junctions and trapping, semiconducting thermistors and rotons in the superfluid helium.

An interesting possibility for a solar neutrino detector was presented by G. Seidel. The principle is based on the calorimetric detection of recoil electrons in elastic $\nu + e^- \rightarrow \nu + e^-$ reactions. Liquid He^4 at low temperatures is proposed as an ideal detection material. To obtain a rate of one event per day a large mass of 10^3 kg is needed. An important feature is that a substantial portion of the electron recoil energy appears as rotons. These are long lived excitations which propagate ballistically at low temperatures and evaporate helium atoms from the free surface of the liquid. The evaporated helium atoms are detected with bolometers mounted above the liquid helium surface.

One of the reports which received great interest at the workshop was given by N. Booth. He and his collaborators have developed a novel indium detector for a solar neutrino experiment. They successfully worked out a method to grow indium crystals over superconducting tunnel junctions. The energy released when a solar neutrino is captured in indium is mostly transformed into phonons. These phonons can break up Cooper pairs in the junction leading to quasiparticles. The tunnel junction acts as a semipermeable membrane letting the quasiparticles pass, thus giving rise to a signal current. The magnitude of this current, however, is very small. N. Booth showed that the sensitivity of the junction could be enhanced by a factor $\sim$100 or more by trapping the quasiparticles in another superconductor of smaller gap in the region of the tunnel junction. If the ratio of the gap energies is larger than 3, a multiplication process can occur leading to a "quasiparticle multiplier".

F. Pröbst, W. Seidel, and F. v. Feilitzsch reported on their work with superconducting tin tunnel junctions, bolometers, and SSG. New developments in very sensitive front-end electronics for bolometers were discussed by D.V. Camin. T. Niinikoski and A. Rijllart described the thermistor approach to detecting coherent neutrino nucleus scattering at high energies. There were also status reports on the dark matter problem, double beta decay, as well as on the recent supernova explosion (E. Fiorini and L. Stodolsky).

In conclusion, very interesting and new approaches dealing with frontier developments in low temperature and solid state physics were presented at the workshop. Most of these ideas are still under feasibility study. The sensitivity of most detection systems reached so far is still orders of magnitude away from what is finally needed. The expenditure on cryogenics is quite considerable. One of the major problems to be solved in the future is the handling of the background, which is due to natural radioactivity of the surroundings and the detector materials, the cosmic ray background, and the detector specific background. Considerable effort has to go into the development of readout electronics of the various devices. Tunnel junctions and SSG are tiny objects which have to be mass produced and have to operate in a reproducible way in a very large detector. This workshop showed that we are only at the beginning of a long way. There was, however, an enthusiastic atmosphere and the general feeling that the meeting was a success. Next year's workshop will be held at LAPP (Annecy).

München, May 1987 *K. Pretzl · N. Schmitz · L. Stodolsky*

Castle
Ringberg

Contents

Update on Neutrinos, Dark Matter, and Cryogenic Detection

L. Stodolsky

Max-Planck-Institut für Physik und Astrophysik,
Werner-Heisenberg-Institut für Physik, Föhringer Ring 6,
D-8000 München 80, Fed. Rep. of Germany

Our meeting takes place under an auspicious sign, that of the supernova in the large Magellenic cloud. With SN 1987A and the observation of an associated neutrino signal, there is now a second fact in astronomy to be added to the pioneering work with the Cl solar neutrino detector in the Homestake mine.

The state of our present knowledge or ignorance of neutrinos may be summarized in a few statements. First, all neutrino interactions - so far - seem to be in good agreement with the standard electro-weak model. The main open questions about neutrinos have to do with the possible existence of more neutrino types beyond the presently recognized [1] ν_e, ν_μ, ν_τ their possible masses,

$$
\begin{aligned}
m_{\nu_e} &\lesssim 18\ \mathrm{eV} \\
m_{\nu_\mu} &\lesssim 250\ \mathrm{KeV} \\
m_{\nu_\tau} &\lesssim 56\ \mathrm{MeV}
\end{aligned}
\tag{1}
$$

the problem of the nature of the antiparticle, and finally the question of neutrino mixing.

In principle, as we studied a few years ago [2], and as was originally proposed by Zatsepin [3], the long flight time from the supernova could provide sensitive information on the neutrino mass. There are two effects which can be used to check each other. One is that if neutrinos of different masses are detected, then there will be pulses separated in time. The other is that within a pulse belonging to a definite mass, there will be a spread in arrival times due to the dispersion in neutrino velocity. Unfortunately, it does not seem possible to get more out of the supernova than we already knew as far as masses are concerned. Not that many papers aren't being written trying to get all kinds of things out of the supernova, but since it seems that supernova models can allow bursts as long as many seconds, and the observations of Kamiokanda and IMB are on that time scale, all we can really say is that the pulse has not been too spread out by mass effects, giving an upper limit of 30 eV or so for the $\bar{\nu}_e$ mass.

By 'the problem of the nature of the anti-particle' we mean the following: A particle may be different from its antiparticle, like the electron is different from the positron, or it may be its own antiparticle, like the photon. Surprisingly, despite our relatively extensive knowledge of the neutrino, we don't know which of these two possibilities applies to it. This is often refered to as the 'Dirac neutrino' (different from its antiparticle) versus the 'Majorana neutrino' (same as its antiparticle). We always talk of neutrino and antineutrino, but this is only defined operationally. That is, what is emitted in the reaction

$$n \rightarrow p\, e^- \bar{\nu}_e \qquad (2)$$

is called an antineutrino, an antilepton. This is because an electron, classified as a lepton, has also been emitted, and we would like to balance lepton number. By the same token, the neutrino emitted in the anti-reaction to (2) is denominated by a neutrino. This is very reasonable and probably even correct. However, there is another possibility, namely that the two objects are actually just two different spin states of the same particle. This curious sounding possibility arises because of the very special way the weak interactions are constructed with respect to spin. The weak interactions depend very strongly on the spin orientation of the particles, so that instead of really being particle and antiparticle ('Dirac') it could be that the neutrino and antineutrino are actually just the same particle, but emitted with different helicities ('Majorana'). (Helicity is defined as the projection of the spin on the direction of motion.) If neutrinos are strictly massless, the helicity could be strictly conserved for neutrinos and could act like a conserved quantum number: lepton number, for example. It thus turns out that with the usual way in which weak interactions, that is, neutrino interactions are constructed, a very light and thus practically always relativistic Majorana neutrino acts the same as a Dirac particle. Although in the Majorana case there is no true antiparticle, the helicity orientation plays the role of particle - antiparticle. In the limit of exactly zero mass, the two options become indistinguishable. It has been determined experimentally that (what we call) neutrinos come out left handed so that in the Majorana case we have left-handed helicity for the so called 'neutrino' and right-handed helicity for the so called 'antineutrino'. The situation if the neutrino has a small mass is then as follows: In the Majorana case we have only two states, helicity + and -. In the Dirac case we have four states: The particle with helicity + and -, and the antiparticle with helicity + and -. However, two of these four states, the right-handed particle and the left-handed antiparticle are practically decoupled from everything (only coupled to order m/E). In the Majorana case, then, we have not true conserved lepton number. We can have interesting phenomena,to the extent that the small m/E effects come in. One is neutrinoless double beta decay:

$$(Z,A) \longrightarrow (Z+2,A) + 2e^- \qquad (3)$$

where only leptons, but no antileptons, come out: lepton number is manifestly violated. So far this process, despite impressive progress in low counting experiments, has not been seen. So either the neutrino is not Majorana or, if it is, its mass is so small that the rate cannot be detected yet. We will be hearing more about this later in the meeting.

The other main open question concerning the neutrino itself has to do with the possibility of neutrino mixing. This possibility arises if neutrinos have non-zero masses. It is then possible that the objects emitted in definite reactions like (2) or say

$$\pi \to \mu \, \nu_\mu \qquad (4)$$

and therefore called ν_e and ν_μ respectively, are not states of definite mass but rather quantum mechanical linear combinations of states having different masses. Then, something emitted as a ν_e or a ν_μ is not in a constant state as a function of time, evolving instead through various linear combinations of different neutrino types. Reactor experiments over the recent years have set good limits on possible mixing, saying essentially that, if mixing exists, either the mass differences must be small [4], $\Delta m^2 = m^2_{\nu_e} - m^2_{\nu_x} \lesssim 2 \cdot 10^{-2} eV^2$ or that the 'mixing angle' characterizing the linear combination must be small, $\sin^2 2\theta \lesssim 0.1$. Since we really have no a priori idea of what these parameters should be, nothing prevents them from being quite small.

This brings us to our next point, one where indeed small parameters arise. The presence of matter can effect mixing. In certain scenarios it can have very interesting consequences, and has been used as an explanation of the 'solar neutrino puzzle'. This possibility, the 'Michaeev-Smirnoff effect' [5], has been much discussed recently. The essential idea is the following: If we could turn the ν_e's produced by nuclear reactions in the center of the sun into another type of neutrino, they would not interact in the Cl detector, and we could explain the absence of the theoretically predicted solar neutrinos. This is an old idea, but it seemed difficult to do it in a natural and convincing way. Either it was hard to get sufficiently large conversion of the ν_e, or one had to assume very special values for the mixing parameters.

By the Michaeev-Smirnoff mechanism, however, it is possible over a broad range of parameters to get complete conversion of the ν_e as they pass through the sun. This magic works as follows: Due to the fact that ν_e has charged current reactions on the electrons in the sun while, say ν_μ does not, ν_e and ν_μ will have different indices of refraction in the solar material. All the necessary scattering amplitudes are known in the electro-weak standard model, so the indices of refraction are known, and are on the order of $1+G\rho/p$, (G, ρ, p, being the Fermi constant, the particle number density, and the neutrino momentum). Now, an index of refraction has in a certain sense a similar effect as a mass for the neutrino. If ν_e and ν_μ have masses and these masses are slightly different, it is possible to arrange for the index of refraction effects and the neutrino masses to compensate so that in effect the two neutrino types are completely degenerate. For a given energy and ρ the two types would have exactly the same wavelength. When this happens ('resonance'), the degeneracy of the two types means that even a very small mixing force can rotate the ν_e completely into ν_μ. At first, this would seem to require very 'fine tuning' to get the indices of refraction just right to cancel the mass difference. Roughly speaking, the cancellation condition is

$$G\rho \simeq \frac{\Delta m^2}{2p} \tag{5}$$

Note the proportionality to the density of material. Along its flight path from the center of the sun the neutrino passes through a wide range of material densities. Thus, the neutrino masses need only be somewhere in a rather wide range for the effect to occur someplace along the way - no 'fine tuning'.

If the density is not changing too fast at this point ('adiabatic condition'), the conversion can be complete. This is analogous to the way a magnetic moment, say a neutron spin, can be completely turned around by a slowly varying field, considering the 'flavor isospin' in analogy to the neutron spin. For more on this see my lecture at the Erice Astro-Particle School [1]. Observe the (neutrino) momentum dependence on the rhs of eq. (5). It is easier to cancel the mass for higher energy neutrinos. It is, therefore, possible to get rid of, say only the high energy ($E > 5$ MeV) solar neutrinos, as for the Cl detector [6], or maybe to get rid of all neutrinos above some lower energy, or maybe practically any solar neutrinos. It turns out that for the mechanism to work we must have ν_μ heavier than ν_e, and typical parameters resulting from eq.(5) are $\Delta m^2 \sim 10^{-5} eV^2$, $\sin^2 2\theta \sim 10^{-3}$, although these parameters can vary widely since the only essential constraints at present are the Homestake numbers [7].

If this explanation turns out to be right, it will be an impressive example of how astrophysical arguments can provide information on extremely small numbers, of the greatest interest to fundamental theory, that would never be accessible in the laboratory.

How can we find out if it is right? Here is an important point for our meeting. The neutral current scattering process we will be discussing in connection with the nuclear recoil detection method is flavor blind and also particle - antiparticle blind, so all neutrinos (so far known) give the same rate in the nuclear recoil detector. Thus, if as just discussed, the missing rate for solar neutrinos in the Cl detector is due to conversion of ν_e into ν_μ or ν_τ on the way to the detector, the neutral current detector should still show the theoretically predicted rate. It would not matter if a ν_e has been converted into another (conventional) neutrino. A comparison of rates in neutral current and inverse beta decay detectors would then give very interesting and sensitive information about neutrino mixings and masses. This is what prompted Weinberg in his summary talk at the Rochester conference in Berkeley last summer to say that neutral current solar neutrino experiments are as important as any now going on in physics.

Following the supernova and the Michaeev-Smirnoff effect we come to the third big development of interest to us. Last and certainly not least, this is the dark matter problem. Inspired by our suggestion [8] that the small energies involved in the nuclear recoil process might be detectable by cryogenic methods, Witten and Goodman [9] pointed out in 1984 that in some models of the dark matter a detectable event rate would arise in the detector due to the particles making up the dark matter. This has understandably created great excitement in the astro-cosmological community - usually in inverse proportion to the distance of the author from the practical problems. Nevertheless, it is undeniably a fascinating possibility and a motivation for us to work harder.

The dark matter problem arises in various contexts, but is most clearly seen in the rotation curves [10] for spiral galaxies. At large distances from the center of the galaxies, both radio (HI) and optical observations match up and show that the velocity of the material orbiting the galaxy is constant as a funtion of distance from the center. This includes distances well beyond where luminous material can be detected and is in marked contrast to what would be expected with a well-defined central mass where we should have the Keplerian $v^2 \sim 1/R$. All kinds of explanations have been contemplated, even changing the law of gravity. If we leave this aside, genuine non-luminous matter in some form seems necessarily to exist, contributing perhaps 90% of the matter in the galaxy. Furthermore, there is evidence for dark matter in other systems such as clusters of galaxies and spherical clusters [11]. One possibility is that it is essentially ordinary matter which does not produce light like burned-out stars or large planets. Apparently such objects are not easy to produce in the present models of galactic evolution, so many people have sought the dark matter in entirely new forms of matter. On the other hand, galaxy formation does not seem that well understood in any case, so conventional matter is by no means totally ruled out.

The possibility that concerns us here, however, is that the dark matter is indeed some new kind of elementary particle(s), presumably relics of the big bang, forming a large spherical halo around the galaxy. From the rotation curves and a virilization assumption, one obtains a density of ~ 0.4 GeV/cm^3 [12] in our vicinity and a velocity typical of most objects in the galaxy, $v/c \sim 10^{-3}$. The particle is weakly interacting, otherwise it could lose its energy and collapses down with the ordinary matter. Some candidate objects, such as the lightest supersymmetry particle, would resemble neutrinos. The sneutrino, for example, would interact via Z° exchange and scatter coherently from nuclei just like neutrinos. The cross section would be governed by a formula similar to that used for the elastic neutrino nucleus scattering which I recall is [8]

$$\sigma \simeq \frac{G^2}{4\pi} [N]^2 E_\nu^2 \qquad (6)$$

Surprisingly, the rates in the detector which result can be quite high, given the assumed density and velocity of the dark matter. The rate estimates from [9] for various assumptions on the mass of the sneutrino can be 1000 kg/day, exceeding that to be expected in high flux neutrino experiments.

This at first sight puzzlingly high rate fact has to do with the E_ν^2 in eq.(6). If we have something like a slow heavy neutrino it turns out the E_ν^2 factor should be interpreted as the full relativistic energy, and so $E \longrightarrow M$. If we are talking about masses like 10 GeV or more, this factor then implies a great increase compared to the MeV we put in for reactor or solar neutrinos. In addition, the kinematics of the collision is different from that with a relativistic neutrino, so that larger energy transfers to the nucleus result. This is because a $v/c = 10^{-3}$ object with M = 50 GeV has a momentum of 50 MeV which it can transfer to a heavy nucleus, while a 1 MeV energy light neutrino has a momentum of 1 MeV, and the energy transfer goes as the square of the momentum transfer.

These high rates and recoils lead, in the case of the heavier hypothetical masses, to a situation where signals should have already been seen in the well shielded semiconductor detectors in operation for double beta decay. This was pointed out by Avignone et al.[13] who used their data to exclude coherently Fermi strength interacting particles with M greater than 15 - 20 GeV. Naturally, there are many assumptions in all this, but it is interesting that it is already becoming possible to say something concrete.

The 'weak charge squared' factor, the bracket in eq. (6), will differ from that for ordinary neutrinos depending on the couplings of the proposed dark matter particle. In general, for coherently interacting objects we might expect something involving the total number of neutrons and protons squared.

An important class of candidates, however, will actually have zero coherent couplings. These are the 'Majorana' neutrinos discussed above, with standard model interactions. A Majorana neutrino is its own antiparticle and thus being truly neutral can have no charge, even in the generalized sense of something like a 'weak charge'. However, as discussed above, when the Majorana particle is moving relativistically, the helicity can play the role of the charge, so that the Majorana and Dirac neutrino can hardly be distinguished. In the dark matter problem, however, with a massive Majorana neutrino, we presumably have to do with the nonrelativistic situation $v/c \sim 10^{-3}$. Thus, we have a case where the Dirac-Majorana difference becomes important; the particle is slow and there is nothing that acts like a charge, and we have no coherent scattering (to order v/c). In this case, we must turn to some other interactions of the object to detect it. The photino of supersymmetry is somewhat different. Being a kind of photon it can interact other than by gauge boson exchange and might have coherent interactions.

For the photino, Goodman and Witten discussed spin-flip interactions with the nucleus. For the non-coherent interaction, depending on the details of the nuclear physics, the rate is roughly like that on a single particle, and there is a substantial reduction compared to the coherent case. The kinematics of elastic scattering, however, is the same whether the particle scatters coherently or not. So for the heavy Majorana case we also expect large recoils compared to those for ordinary neutrinos. For the spin-coupled particles one may also consider nuclear excitation as well as elastic scattering. The fact that we expect a strong difference between nuclei with spin zero and nuclei with spin could be useful in establishing an effect. A further point noted by Drukier, Freese and Spergel [12] is that there should be yearly variation of the counting rate. The motion of the earth around the sun (30 km/sec) either adds or subtracts to the velocity (250 km/sec) of the sun relative to the galactic center, altering the flux of dark matter seen by the detector and, therefore, the rate. The maximum should be around the end of April. This also could ultimately prove to be important in separating an effect from background.

In conclusion, it is clear that there is no lack of ideas and speculations about what to do with our instruments once we have got them. I am sure that in the course of the meeting it will also become clear that there is no lack of ideas and speculations about cryogenic devices. Now, all we have to do is make them work.

References

1. For a review of neutrino properties see the talk by DiLella, Astro-Particle School, Erice, Jan. 1987
2. P. Reinartz and L. Stodolsky: Z. Phys. C27, 507 (1985)
3. G.T. Zatsepin, ZhTEF Pis Red. 8, 333 (1968)
4. G. Zacek et al.: Phys. Rev. D34, 2621 (1986)
5. S.P. Michaeev and A. Yu. Smirnoff, talk given at the 10th Intern. Workshop on Weak Interactions, Savonlinna, Finland, June 16-22, 1985, Yadernaja Fizika (Sov. J.), 42, 1441 (1985), Nuovo Cim. C9, 17 (1986)
6. H.A. Bethe, Phys. Rev. Lett. 56, 1305 (1986)
7. For a study of various possibilities see W. Haxton, Phys. Rev. D35, 2352 (1987)
8. A. Drukier and L. Stodolsky, Phys. Rev. D30, 2295 (1984), MPI preprint PAE/PTh 36/82 (1982)
9. M.W. Goodman and E. Witten, Phys. Rev. D31, 3059 (1984); see also Ira Wasserman, Phys. Rev. D33, 2071 (1986)

10. See the talk by R. Sancisi (ref. 1)

11. A general introduction to these questions is in the Scientific American supplement 'Galaxies'

12. A.K. Drukier, K. Freese and D.H. Spergel, Phys. Rev. D33, 3495 (1986)

13. Ahlen, Avignone, et al.: Limits on Cold Dark Matter from the Ultra-Low Background Germanium Spectrometer; CFA preprint
see also G. Gelmini, Quarks and Galaxies Workshop, Berkeley, July 1986

New Results on the Basic Properties of Superheated Granules Detectors

L. Gonzales-Mestres and **D. Perret-Gallix**

L.A.P.P., B.P. 909, F-74019 Annecy-le-Vieux Cedex, France

Abstract: Recent results on the basic physics of superheated superconducting granules (SSG) detectors are presented.

Real time detection of single granule flips under individual particle interaction is reported and studied in detail. Irradiation of very large tin granules ($45\,\mu m \leq \phi$ diameter $\leq 400\,\mu m$) with α particles (^{241}Am, $E \simeq 5.5$ MeV) shows evidence for local heating phenomena, where the observed energy threshold is far below that predicted by equilibrium thermodynamical calculations. Preliminary results on irradiation of $200\,\mu m \leq \phi \leq 400\,\mu m$ grains with 6 keV γ's (^{55}Fe) are also briefly discussed.

Using smaller granules ($10\,\mu m \leq \phi \leq 25\,\mu m$), a systematic irradiation study with γ and β^- sources has been carried out: ^{99}Tc 140 keV γ's, ^{36}Cl $E \leq 714$ keV β^-, ^{55}Fe 6 keV γ's. In all cases, flips have been observed and the sensitivity may be better than that predicted by the global heating model. About 8% of the granules turn out to be sensitive to 140 keV γ's.

Tests made at lower temperatures ($T \geq 450$ mK) show the absence of avalanche effect (seen by other authors in different conditions) for several samples of tin granules. A theoretical discussion of the avalanche effect is then presented. Estimates of the behaviour of the detector at very low T are given, where a thin layer of normal electrons near the surface is shown to contribute to the heat capacity of a superheated granule.

Consequences for some of the proposed experiments are sketched. Prospects for the development of SSG detectors in the near future are discussed, leading to small size (1 cm^3) three-dimensional multichannel prototypes, where sensitivity and energy resolution can be better studied.

1. INTRODUCTION

The idea of using superheated superconducting granules [1] for particle detection [2] is due to the Orsay group [3] and the first series of tests with electron beams was made in Rennes [4]. X-ray imaging [5] or transition radiation detection [6] led to the first attempts to build prototypes of photon detectors. More recently, the growing interest in high sensitivity devices and cryogenic techniques for Astrophysics and Particle Physics has triggered new developments in the field.

Solar neutrinos can possibly be detected with SSG detectors, either through the nucleus recoil energy [7] or, using indium granules [8], through Raghavan's reaction [9]. Magnetic monopoles are expected [10, 11] to destroy the superheated state independently of the size of grains used. Thus, a large area detector may be feasible if large granules of good quality are available (good background rejection, better signal over noise ratio). Cold dark matter [12, 13, 14] and solar axions [15] may be detectable with suitable materials such as Al or Ga, if sensitivity to energies below 1 keV is demonstrated. Finally, double β experiments [16] could be performed with better space resolution (but much worse energy resolution) with cadmium granules, where ^{116}Cd (7.58%) is a $\beta^-\beta^-$ nucleus with Q = 2.811 MeV and ^{114}Cd leads to $\beta^-\beta^-$ with Q = 547 keV. Molybdenum granules (if available, and if they can reach the superheated state) would be a most interesting material because of the high value of Q (3.03 MeV for ^{100}Mo).

The widespread range of applications proposed makes it important to emphasize that SSG detectors are still at the level of a feasibility study. Present experimental tests are at the moment limited by the difficulty to produce very small granules ($\phi < 5\ \mu$m) with the required size homogeneity. But there are also obstacles related to the basic properties of the detector. One major drawback is the spreading of the superheating curve, i.e., the dispersion in the effective superheated critical field H_{sh}^{eff} seen by each granule. This implies a loss in efficiency for a required sensitivity. For single grains, H_{sh}^{eff} has been shown [17] to depend smoothly on the orientation of the granule with respect to the applied magnetic field. Another difficulty may be the existence of an avalanche effect [18] that, for some materials and with certain cooling techniques, seems to prevent from operating the detector at temperatures far below T_c (the critical temperature), where the fast decrease of superconducting specific heats would provide higher sensitivity. Finally, in some cases the use of very small granules may lead to signals beyond the reach of present day electronics, especially if good time resolution is simultaneously required.

It therefore seems that the present situation demands:

1. To improve the technique of grain production at the industrial level, in order to obtain collections of good spheres without surface defects, providing a narrow superheating spectrum. A better homogeneity in size would also be a leap forward, and may become a crucial issue in the fabrication of small grains.

2. To better understand the basic physics of the detector, in order to search for solutions to the many yet unsolved problems. This is the goal of the present work.

Chapter 2 presents experimental results on the irradiation of very large granules (up to 400 μm diameter) with α particles, allowing for the first time to obtain quantitative information on the mechanism of phase transition by a local deposit of energy. The interpretation of these results, as well as possible consequences, are discussed in chapter 3, where we claim evidence for a local heating mechanism in which nucleation of the normal state occurs before the granule is thermalized. Preliminary results on the irradiation of $200\ \mu m < \phi < 400\ \mu m$ with 6 keV photons provide a dramatic confirmation of our analysis. Chapter 4 presents irradiation results with smaller granules ($10\ \mu m < \phi < 25\ \mu m$) using γ and β^- sources. In all cases, flips have been seen. A résumé of our data is given, including ^{55}Fe γ irradiations.

Chapter 5 presents results and estimates on the behaviour of the detector at very low temperature, in particular: a) The observed absence of avalanche effect for several samples of Sn granules at $T \geq 450$ mK; b) The heat capacity of small superconducting spheres in the presence of an applied magnetic field, where normal electrons near the surface turn out to play an important role; c) Numerical estimates concerning some of the proposed experiments. Conclusion and comments, including prospects on SSG development, are presented in chapter 5.

The concepts and notations used throughout this contribution are the same as those of reference [2], where a general introduction to the subject can be found.

2. EXPERIMENTAL RESULTS WITH α PARTICLES

Our results on α irradiation were presented in a preliminary form at the previous München Meeting last year. We give, here, a brief summary of our main conclusions.

Recent progress in grain fabrication allowed us to obtain, from industrial producers, collections of very large granules exhibiting good metastability properties. Then, for the first time, the mechanism of phase transition from metastable to normal state could be studied in detail.

The following collections of tin granules were used:

Sample a_1): grains of size $45\ \mu m \leq \phi$ (diameter) $\leq 63\ \mu m$, with normal state resistivity $\rho \simeq 6 \times 10^{-8}\ \Omega$.cm (as measured before grain fabrication).

Sample a_2): grains of size $125\ \mu m \leq \phi \leq 200\ \mu m$ made of the same material.

Sample b): grains of size $250\ \mu m \leq \phi \leq 300\ \mu m$, $\rho \simeq 1.5 \times 10^{-7}\ \Omega$.cm.

Sample c): grains of size $200\ \mu m \leq \phi \leq 400\ \mu m$, made of an alloy $Sn_{99}Sb_1$, ρ(estimated) $\approx 10^{-6}\ \Omega$.cm

Collections a_1), a_2) and b) were produced by EXTRAMET [19], whereas collection c) was prepared by BILLITON [20]. Collections a_1) and a_2) were separated by sieving from a larger collection ($45\ \mu m < \phi < 200\ \mu m$). The samples were prepared embedding the granules into paraffin with dilution coefficients in the range of 5-20% in volume.

All samples were put in contact with an open α (5.5 MeV) source (^{241}Am) with a 4π activity of 5000 Bq. As a consequence, a few ^{241}Am atoms drifted into the samples, which remained irradiated even after the source was removed. The number of contaminated granules turned out to be small as compared to the total number of granules in the sample. The tests were performed at temperatures between $T \simeq 1.4$ K and the tin critical temperature ($T \simeq 3.7$ K). Detection loops were multi-turn (6 to 15) coils (6 mm to 1 cm diameter) made of thin copper wires (100 to 200 μm in diameter). The coil was connected to the pre-amplifier through a 1.5 m twisted pair transfer line ($\simeq 1$ μH). The coil parameters were optimized to match the transfer line inductance. At low temperature the resistance of the system was of the order of 1 Ω.

Several read-out systems have been used, since the signals turned out to be detectable with standard low noise amplifiers. Risetimes of less than 1 μsec were seen for all samples at the output of a 142B Ortec pre-amplifier. The magnetic field was created by a pair of Helmholtz coils installed outside the cryostat (a 10 A current was required to create a 260 G field at the center of the system).

A CAMAC data acquisition system has been set up and automatic procedures have been developed to drive the magnetic field, record temperature and magnetic field and plot hystheresis and long term flip rate curves. Fig. 1 presents supercooling and superheating curves for sample a_2). As a general trend, the ratio H_{sh}/H_{sc} tends to decrease as the size of granules and the residual normal state resistivity increase.

For sample c), the supercooling curve was found to be lower than the superheating one due to the demagnetizing coefficient associated to superconducting spheres. One has:

$$H = 3/2\, H_o \sin\theta \quad \{1\}$$

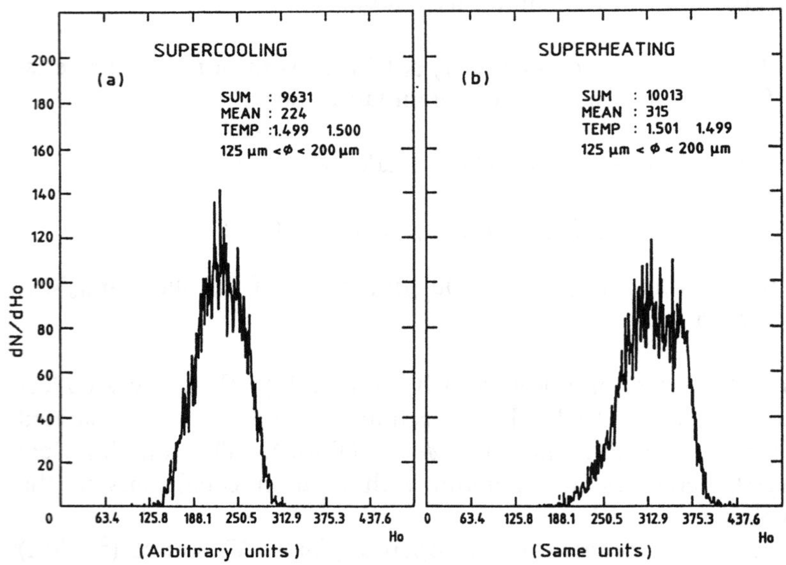

Fig. 1: Superheating and supercooling curves for a collection of 125 μm $< \phi <$ 200 μm EXTRAMET tin granules (sample a_2)

where $\vec{H}_0$ is the applied magnetic field, H the value of the external magnetic field on the surface of the granule and θ the angle between $\vec{H}_0$, and the radius $\vec{R}$. The fact that, at the equator ($\theta = \pi/2$), $H = 3/2\, H_0$, explains that the effective superheated critical field H_{sh}^{eff} may be lower than supercooling. In such case, we are confronted to transitions from normal or superconducting state to the intermediate state.

Measurements were performed in the following way. The applied magnetic field H_0 is slowly raised from 0 to a certain value H_{test} (usually close to the maximum of the superheating curve). At this point a small sweep in H_0 is performed, establishing a gap in magnetic field ΔH_{min}. An energy threshold is therefore created for the remaining superconducting grains. Granules flipping for $H_{test} < H_0 < H_{test} + \Delta H_{min}$ are no longer in the superheated state and therefore are not sensitive to α particles. Those changing state above $H_{test} + \Delta H_{min}$, must receive enough energy to break an energy barrier corresponding to $\Delta H > \Delta H_{min}$. In the rest of this chapter, H is the value of H_0 for which a given granule actually flips, and $\Delta H = H - H_0$. This can be seen on Fig. 2, where a typical dependence of H_{sh}^{eff} versus T is presented. The threshold ΔH is then equivalent to a threshold in temperature, ΔT. An electronic threshold on the pulse height was adjusted largely above the pedestal. The rate of counts was found to be compatible with zero for non-irradiated samples. Irradiation results can be presented in terms of dN/dt (number of counts per unit time), or through the integrated number of counts N(t). A useful parameter to characterize irradiation curves is:

$$(\Delta H/H)_{min} = \Delta H_{min}/(H_{test} + \Delta H_{min}) \qquad \{2\}$$

Fig. 3 shows the integrated number of counts as a function of time for a 5 hours irradiation test with granules of collection a_1) at $T \simeq 1.5$ K and $(\Delta H/H)_{min} \simeq 0.013$ $(0.013 \leq \Delta H/H < 0.178)$. The rough exponential behaviour of the slope clearly indicates a decay phenomenon, where the count rate decreases

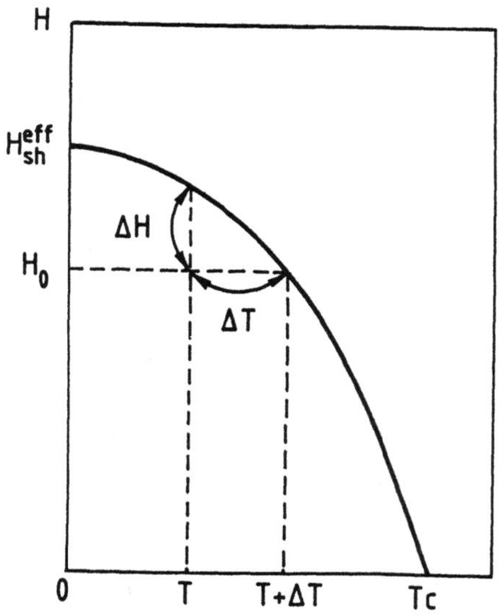

Fig. 2: At fixed H_0, ΔT is obtained from ΔH using the T dependence of H_{sh}^{eff}.

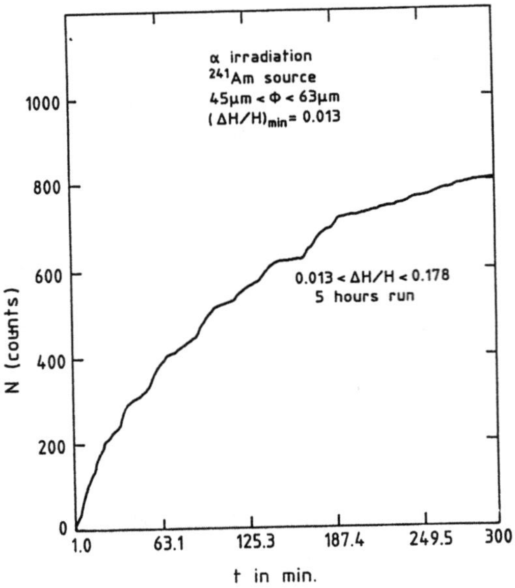

Fig. 3: Integral irradiation curve for sample a_1 (5 hours test), in the presence of a ^{241}Am source (5.5 MeV α's).

as the irradiated granules change state. We performed similar tests for all the above mentioned sets of granules, for values of $(\Delta H/H)_{min}$ between 0 and 0.2. In all cases, flips were observed even for the highest values of ΔH_{min}. As $(\Delta H/H)_{min}$ was set higher, the number of sensitive granules was seen to decrease and the variation of dN/dt with time became slower. The fact that granules as large as 400 μm in diameter were sensitive to 5.5 MeV α particles at $(\Delta H/H)_{min} \simeq 0.2$ provides amazing evidence for a local heating mechanism, as will be discussed in the next chapter.

The observed integral irradiation curves could in all cases be fitted by an expression of the type:

$$N(t) = \int_{A_1}^{A_2} \Psi(A)\,(1 - \exp(-A\,t))\;dA \qquad \{3\}$$

Where $\Psi(A)$ is a differential flipping probability, that was set either linear or quadratic.

3. GLOBAL VERSUS LOCAL HEATING MODEL

Most of the predictions found in the literature concerning SSG sensitivity assume a global heating model, where heat first spreads to the whole granule and subsequently the superconducting to normal phase transition occurs. Then, the particle must deposit enough energy to raise the temperature of the whole sphere in order to induce a flipping process. From standard thermodynamical calculations, assuming uniform distribution of heat inside the granule, grains of sample a_1) were expected to be sensitive to individual 5.5 Mev α particles for $\Delta H/H < 0.016$ at $T \simeq 1.5$ K. For larger values of $\Delta H/H$, a sharp cutoff would be the normal behaviour assuming global heating to be the relevant mechanism. The fact that flips were observed at much larger values of ΔH_{min} provides clear evidence for a local heating mechanism. At $T \simeq 3.4$ K, the amount of heat ΔQ required for a 50 μm grain flip with $\Delta H/H = 20\%$ in a global heating situation would be an order of magnitude larger than the energy of the α particles. At $T \simeq 1.5$ K, one gets $\Delta Q \simeq 130$ MeV. The calculation of ΔQ is based on the formula:

$$\Delta Q = \int_{T}^{T+\Delta T} C(T')\, dT' \qquad \{4\}$$

where C(T) is the heat capacity of the superconducting granule.

The main result of our tests with α particles is therefore that they rule out the global heating model and provide evidence for a local heating mechanism. In other words, nucleation of the normal state on the surface of the grain seems to happen irreversibly before heat spreads to the whole granule. The possibility of a phase transition by local heating is due to the metastable character of the superheated state, where a nucleation center of the normal state on the surface of the sample is enough to induce the transition of the whole specimen.

A complete model of local heating has never been elaborated, but it mainly implies a precise balance between two time scales. One, the time required to create a permanent nucleation center at the surface. It depends essentially on: a) The value of the external magnetic field H in the region where the deposit of energy takes place. b) The normal state resistivity of the superconductive material. We hereafter call this time scale the "nucleation time" τ_N. The second one is the time required for the heat wave to spread over a distance larger than the critical size of a nucleation center of the normal state. The distance $r(\tau_N)$ reached by heat in a time τ_N depends on the heat diffusion coefficient $D = K/c$ (K = thermal conductivity, c = specific heat), as will be briefly discussed later on. The main difficulty lies in estimating accurately τ_N, in terms of H, ρ and the amount of energy ΔE deposited near the surface of the grain.

The penetration of the magnetic field into superconducting samples at $H > H_c$ has been carefully studied for cylinders [21, 22], and some of these results have been extrapolated to spheres [5] without further justification. We do not believe such an extrapolation to be entirely correct, since superheated spheres can go to the intermediate state, which is not the case for cylinders when the applied field H_o is parallel to the axis. But, locally, this description [21] and [22] is likely to give the correct order of magnitude for a small penetration around a nucleation center on the surface of the sphere. In the simplest model [21], the flipping time of a cylinder of radius R is:

$$\tau \simeq \rho^{-1} \pi R^2 H_o/(H_o - H_c)$$

where H_c is the critical field and the time evolution equation is:

$$t(d_N) \simeq \rho^{-1} \pi H_o/(H_o - H_c) [R^2 - (R - d_N)^2(1 + 2 \ln R/(R - d_N))]$$

and d_N is the distance to which the magnetic field has penetrated. Then, an ansatz for τ_N, to get the right order of magnitude, could be to take $d_N \sim \xi$ (the coherence length) and write:

$$\tau_N \sim 2\rho^{-1}\pi\xi^2 \, H/(H - H_c) \quad \{5\}$$

and H is the value of the field near the point of the surface where the deposit of energy took place. Then, for tin ($\xi \simeq 2300$ Å) and $H \simeq 2 H_c$, we get: $\tau_N \sim 6 \times 10^{-9}$ sec for $\rho \sim 10^{-9}$ Ω.cm and $\tau \sim 6 \times 10^{-12}$ sec for $\rho \sim 10^{-6}$ Ω.cm. These two values correspond to the minimum and maximum normal state resistivity of materials used in existing prototypes. In the last case, we reach the characteristic time scale of Cooper pairs, where microscopic effects are expected to play an important role.

Changing ρ also modifies the value of D, mainly through the electron contribution to the thermal conductivity K. For tin alloys, data at superconducting temperatures [23] show a linear T dependence consistent with the Wiedemann-Franz law, whereas the ratio K(superconducting)/K(normal) remains reasonably close to 1 in the whole range $0.4 < T/T_c < 1$. In what follows, we neglect microscopic phenomena and use the heat diffusion equation :

$$\partial \Delta T / \partial t = D \, \Delta' (\Delta T)$$

where, in the second term, Δ' means "laplacian". A rough estimate of the possibility of local heating phenomena can be given considering the case of an α particle releasing its energy on the surface of the grain, and approximating the distribution in r of ΔT by a uniformly hot half sphere of radius r(t). For r(t) we take the maximum of the distribution $\exp(-r^2/4Dt)$ that appears when considering the three dimensional isotropic diffusion equation. One then gets: $r(t) \simeq 2\sqrt{tD}$. The temperature at the center of the sphere varies as $t^{-3/2}$, as usually in three dimensional diffusion phenomena. To further simplify the calculation, we ignore effects due to the granule-paraffin interface and assume that the above formulae apply to a half sphere of radius $r(\tau_N)$. From data on thermal conductivity [23] and specific heat [24, 25], we can make a rough estimate of the energy required in order to heat a half sphere of radius $r(\tau_N)$ centered on a point of the surface of the grain, and produce nucleation for $\Delta H/H > 0.2$, at $T \simeq 1.5$ K. With the value $\rho \simeq 6 \times 10^{-8}$ Ωcm, one gets:

$$r(\tau_N) \sim 11 \ \mu m \text{ and } \Delta Q \sim 3 \text{ MeV}$$

which is compatible with our experimental results. It should be realized, however, that some crude approximations were involved in the present calculation.

More recently, we have performed an additional test to check the local heating mechanism, using granules of sample c). Due to the high residual normal state resistivity of such grains, local heating at the $\approx 1 \mu m$ scale may be expected. Calculations at the microscopic scale for small In granules [26] confirm such an approximation if the above expression for τ_N is used. The granules were irradiated with ^{55}Fe γ's ($E \simeq 6$ keV), mixing the source directly with the detector. Remark-

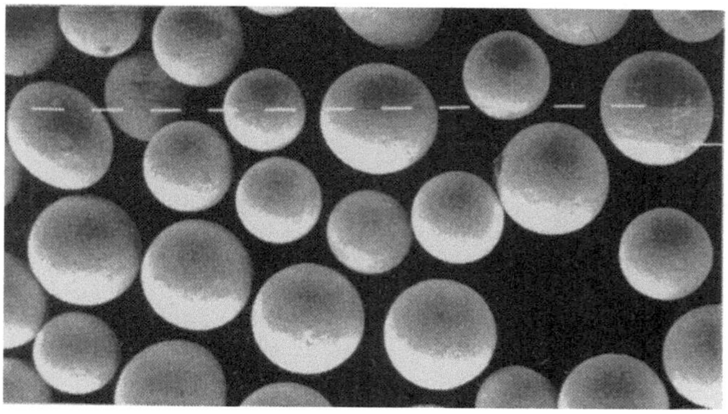

Fig. 4: 200 μm < ϕ < 400 μm BILLITON granules (alloy $Sn_{99}Sb_1$ sample c). The mark is 100 μm.

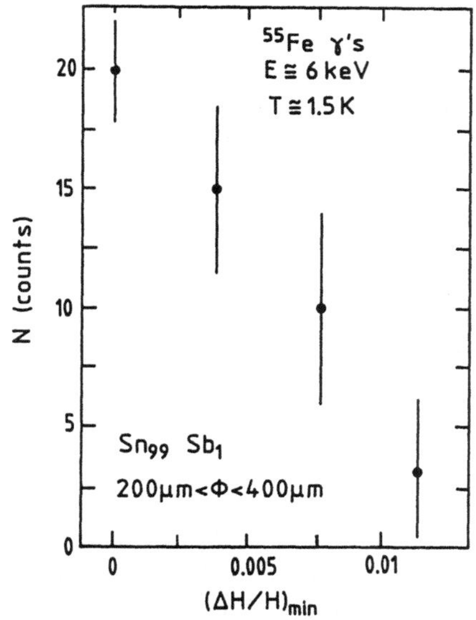

Fig. 5: Counts after 1 hour for sample c) irradiated with 6 keV γ's versus $(\Delta H/H)_{min}$.

ably enough, flips were observed for $(\Delta H/H)_{min} \leq 0.01$. In a global heating model, for ϕ = 200 μm, this would correspond to an energy threshold of about 120 MeV, i.e. four orders of magnitude larger than the energy of ^{55}Fe photons. Such data lead to a $r(\tau_N)$ of less than 5 μm. The irradiated granules are shown in Fig. 4 , whereas Fig. 5 shows the variation of the number of flips seen after 1 hour, as a function of $(\Delta H/H)_{min}$. The dynamics of local heating is certainly a crucial item and deserves further study.

4. IRRADIATION RESULTS WITH γ AND β^- SOURCES.

For γ and β^- irradiations, smaller granules were used:

Sample d). From a large batch of granules produced by EXTRAMET in December 1986, a collection with 10 μm < ϕ < 25 μm was extracted by sieving. A

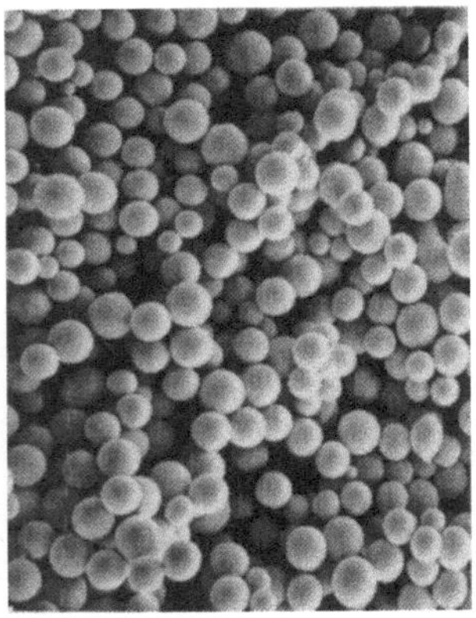

Fig. 6: 10 μm < ϕ < 25 μm granules (sample d). The mark is 10 μm.

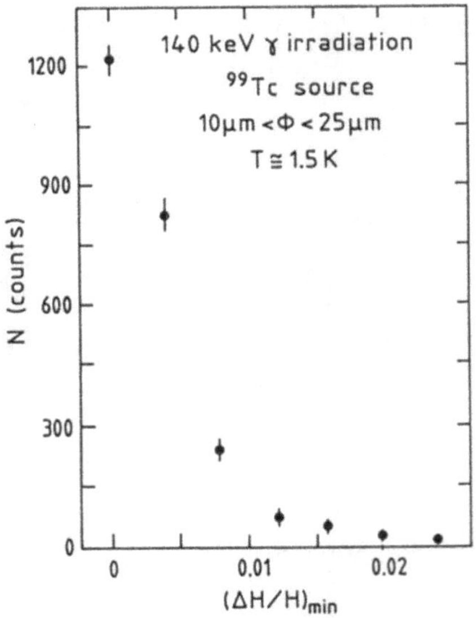

Fig. 7: Counts after 20 min. for sample d) irradiated with 140 keV γ's versus $(\Delta H/H)_{min}$.

photograph of these grains is shown on Fig.6. The observed dispersion in H_{sh}^{eff} was about $\pm 10\%$, and the normal state resistivity of the tin used was about 10^{-7} Ωcm. Electronic read out of such granules was again done in real time with standard low noise amplifiers.

Several sources were used:

- γ irradiations were made with a ^{99}Tc source ($E_\gamma \simeq 140$ keV, $t_{1/2} \simeq 6$ hours) introduced inside the cryostat close to the detector prototype. About 8% of the granules were found to be sensitive to the ^{99}Tc photons, and flips were observed at $(\Delta H/H)_{min}$ larger than 0.02. The dependence of the number of flips (accumulated after 20 min) on the relative threshold in H_0, $(\Delta H/H)_{min}$, is shown in Fig. 7. The fact that the flips were due to individual γ's was checked by a plot of $dN/dt|_{t=0}$ in terms of the source activity [27]. In a global heating model, the value $(\Delta H/H)_{min} \simeq 0.02$ would correspond to $\Delta Q \simeq 60$ keV for grains of $\phi = 10\ \mu$m and $\Delta Q \simeq 1$ MeV for $\phi = 25\ \mu$m.
- β^- irradiations were performed with ^{36}Cl ($E < 714$ keV) and ^{14}C ($E < 158$ keV), mixing the sources directly with the detector. Grain transitions were seen for $(\Delta H/H)_{min} \leq 1\%$. Fig. 8 shows the variation of the number of observed flips after 1 hour, as a function of $(\Delta H/H)_{min}$, for the ^{36}Cl source.
- Finally, we studied the sensitivity of sample d) to very low energy sources (^{3}H, ^{65}Zn, ^{55}Fe). In all cases, flips were seen at small ΔH_{min} We present,

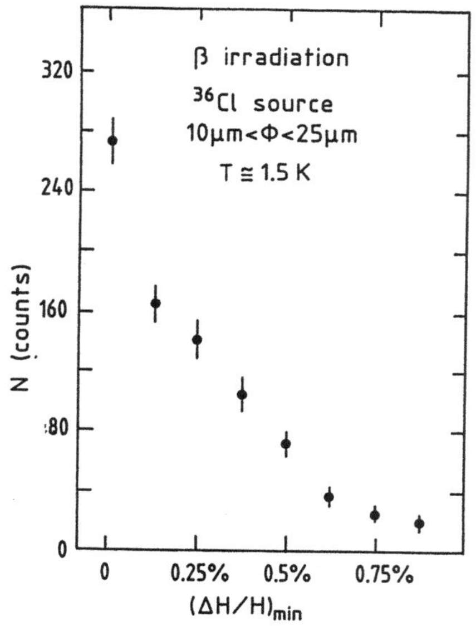

Fig.8

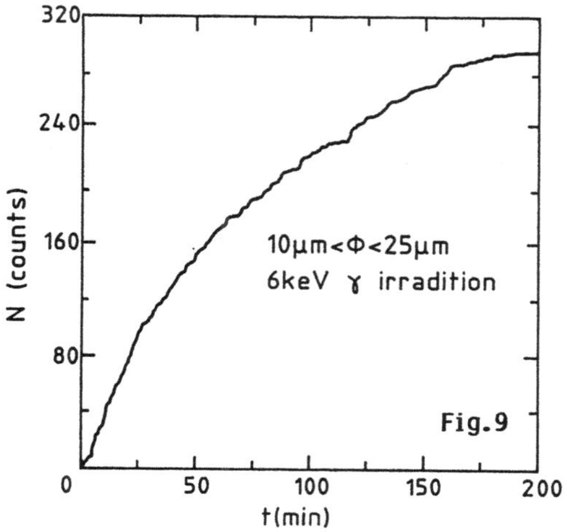

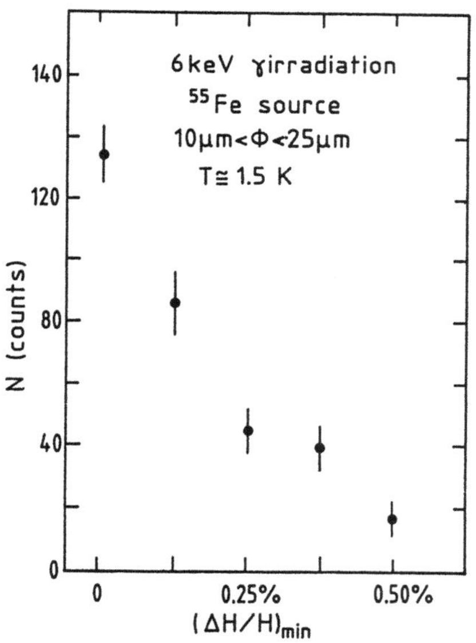

Fig. 8: Counts after 1 hour for sample d) irradiated with E < 714 keV β^- versus $(\Delta H/H)_{min}$.

Fig. 9: Integral irradiation curve for sample d) in the presence of a ^{55}Fe source (6 keV γ's).

◄ Fig. 10: Number of counts after 1 hour for sample d) irradiated with ^{55}Fe γ's.

here, results with 6 keV γ's of ^{55}Fe. Fig. 9 shows an integrated irradiation curve at $(\Delta H)_{min} \simeq 0$ and Fig. 10 the dependence on $(\Delta H/H)_{min}$ of the number of counts accumulated after 1 hour. The observed sensitivity at $(\Delta H/H)_{min} \leq 0.005$ corresponds, in global heating calculations, to $\Delta Q \simeq$ 15 keV for $\phi = 10$ μm, and $\Delta Q \simeq 250$ keV for $\phi = 25$ μm. Again, we cannot exclude the possibility of a local heating mechanism even for the smallest granules.

The results obtained with γ and β^- sources clearly establish the sensitivity of SSG to low energy particles. However, efficiency and energy resolution are at the moment far from those required to build a particle detector. Smaller granules are required, which implies serious improvement in the electronic read out. We are now preparing an irradiation test of $\phi < 10\ \mu m$ Sn granules at $T \geq 300$ mK.

5. PROSPECTS ON VERY LOW TEMPERATURES.

A possible way to improve the performance of the detector is in principle to run it at very low temperature. The specific heat of a superconducting material is given by:

$$c_s = c^{ph} + c_s^{el} \qquad \{6\}$$

where the phonon term varies as:

$$c^{ph} \simeq a\, T^3 \qquad \{7\}$$

with $a \propto \Theta^{-3}$, Θ being the Debye temperature, and the superconducting electron term c_s^{el} is often parametrized [24, 28] by:

$$c_s^{el} \simeq b\, \gamma\, T_c \exp(-f\, t_r^{-1}) \qquad \{8\}$$

$t_r = T/T_c$ is the reduced temperature and γ the normal electron specific heat coefficient, whereas b and f are determined experimentally.

A theoretical discussion of the sensitivity of SSG detectors at $t_r \ll 1$ has been given in [2]. Here we further develop two points:

5.1 THE AVALANCHE EFFECT

According to [18], latent heat released by the transition of a Cd granule produced the flip of other granules, which in turn released more heat so that many other grains changed state leading to an almost complete collapse of the detector at $T \leq 350$ mK ($t_r \simeq 0.6$). In contrast with these data, we performed tests at $T \geq 450$ mK with several samples of tin granules, at about 10% filling factor in volume, and no avalanche effect was found. It is therefore important to understand whether such a difference is related to the superconductive materials used, or it should be attributed to the properties of two different cryogenic set-up.

In equilibrium, superconducting to normal phase transition absorbes latent heat [29]. In the case of superheating, there is an extra term coming from the difference in free energy between normal and superconducting state. At $H_o > H_c$ the phase transition goes from superconducting to normal state and one has:

$$\Delta Q^L = F - A_s(H_c)\, V + T\, V\, d/dT\, (A_n - A_s) \qquad \{9\}$$

where ΔQ^L is the released heat, F the free energy of the superconducting sphere in the presence of a magnetic field $\vec{H}_o$, V the volume of the granule, A_s the free energy per unit volume of the superconducting state as referred to the same state in the absence of magnetic field, and A_n the free energy density of the normal state:

$$A_s = H^2/8\pi \qquad \{10\}$$

$$A_n = H_c^2/8\pi \qquad \{11\}$$

For $2/3\ H_c < H_o < H_c$, one has to deal with a transition from superconducting to intermediate state, where:

$$\Delta Q^L = F - F_i + T\ d/dT\ (F_i - F) \qquad \{12\}$$

where F_i is the free energy density in the intermediate state:

$$F_i = V/8\pi\ (H_c^2 - 3(H_c - H_o)^2) \qquad \{13\}$$

The final expression for ΔQ^L in units of $H_c^2/8\pi$ has been given in [2, 27]. We study here more closely the condition for an avalanche to occur. In what follows we assume uniform heat propagation inside the detector, where only the granules and the dielectric material are present (detector in vacuum). The situation may be more complex in the presence of superfluid helium, which is a good thermalizer.

Assume that, at some fixed T and H_o, an avalanche propagates carrying an increase in temperature ΔT and therefore flipping granules at $\Delta H \le \Delta H(\Delta T)$. We can then write a consistency condition requiring that the heat released by granule flips is enough to produce the above increase in temperature:

$$V^{-1}\ \Delta Q^L\ C_{det}^{-1}\ \Delta\psi = \Delta T \qquad \{14\}$$

where ψ is the filling factor in volume and $\Delta\psi$ the filling factor for the fraction of granules flipping at $\Delta H \le \Delta H(\Delta T)$. C_{det} is the average heat capacity per unit volume of the detector. On the other hand, $\Delta H(\Delta T)$ is related to ΔT through the expression:

$$H_{sh}(T + \Delta T)/H_{sh}(T) = H_o/(H_o + \Delta H) \qquad \{15\}$$

In the limit $\Delta H, \Delta T \ll 1$, we get from the last two equations:

$$\delta \equiv -\ V^{-1}\ \Delta Q^L\ C_{det}^{-1}\ T_c^{-1}\ H_o\ d\psi/dH_o\ H_{sh}^{-1}\ dH_{sh}/dt_r = 1 \qquad \{16\}$$

and the avalanche effect is expected to happen for $\delta > 1$. If $\delta < 1$, the flip of a granule can still generate other flips but the chain will stop before reaching macroscopic size. From {16} we can compute, for a given detector, the temperatures and values of H_o at which the avalanche effect can be expected to occur. ΔQ^L can be written [27, 2] as:

$$\Delta Q^L = \Delta Q^L{}_r \, V \, H_c{}^2/8\pi \qquad \{17\}$$

where $\Delta Q^L{}_r$ is a function of $h = H_o/H_c$ and t_r. Assuming $C_{det} \simeq A\,T^3$, and taking [2]:

$$H_{sh}(t_r) \simeq H_{sh}(0)\,(1+t_r{}^2)^{5/12}\,(1-t_r{}^2)^{11/12} \qquad \{18\}$$

derived from approximations based on the Extreme Anomalous Limit [30], equation {16} becomes:

$$\delta \simeq A^{-1}\,\Delta Q^L{}_r\, t_r{}^{-2}\,(1 + 8/3\, t_r{}^2)(1-t_r{}^2)(1+t_r{}^2)^{-1}\, H_o\, d\psi/dH_o\, H_c(0)^2/8\pi T_c{}^4 \qquad \{19\}$$

Making the rough assumption that the dielectric dominates the heat capacity of the detector, we can take A to be independent of ψ, and of the superconducting material used. Then, at fixed t_r and h, the dependence on the superconducting material appears through the factor $\zeta = H_c(0)^2/8\pi T_c{}^4$. For Sn, one has: $\zeta = 19.3$ erg cm^{-3} K^{-4}. In the same units, we get: $\zeta = 25.6$ for In, 73.3 for Ga and 364.1 for Cd. Such figures clearly show that the avalanche effect should occur much more easily for Cd than for Sn. Also, since superheating is lower for Sn than for Cd, $\Delta Q^L{}_r$ is smaller enhancing therefore the difference in behaviour of the two materials. We have attempted a rough numerical estimate for Cd, Sn, In and Ga, setting: $A \simeq 4 \times 10^{-5}$ joule cm^{-3} K^{-4}, $\psi = 10\%$ for Sn, In and Ga, and 4% for Cd., and taking the top of the superheating curve at $H_o = H_c$ in the case of tin, $H_o = 1.6\, H_c$ for cadmium and gallium, $H_o = 1.12\, H_c$ for In. The superheating curve has been approximated by a gaussian distribution of width $2\sigma = 0.16$ for

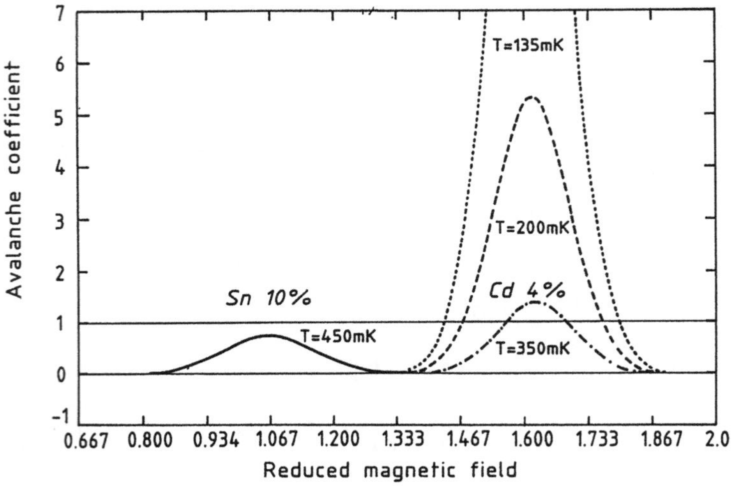

Fig. 11a: Avalanche coefficient δ in terms of the reduced magnetic field $h = H_o/H_c$, for: 1) Sn granules at T = 450 mK and filling factor (volume) ψ = 10%, 2) Cd at ψ = 4% and T = 350 mK (dot−dash line), T = 200 mK (dashed line), T = 135 mK (dotted line).

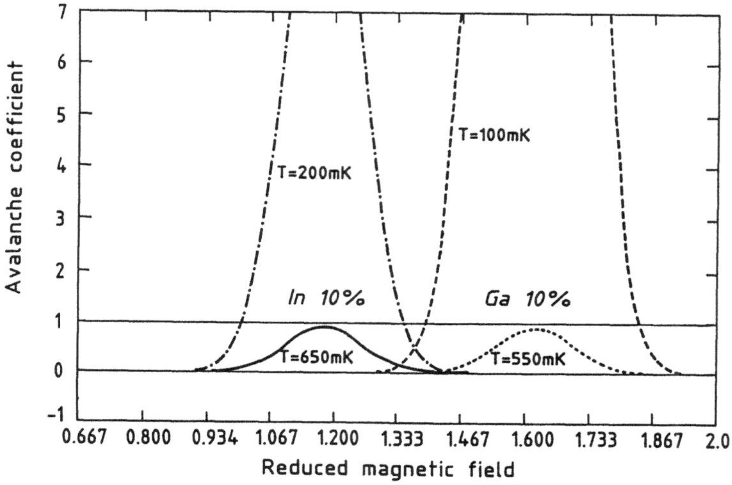

Fig. 11b: Avalanche coefficient δ in terms of reduced magnetic field h for: 1) In at ψ = 10% , and T= 650 mK (full line), T = 200 mK (dot–dash line), 2) Ga at ψ = 10%, T = 100 mK (dashed line), T = 550 mK (dotted line).

Cd, In and Ga, and 2σ = 0.2 for Sn. The numerical results for several temperatures are shown in Fig. 11, where δ is presented as a function of h. The temperatures choosen for Cd correspond to those of superheating curves reported by the Garching group [18], where avalanches were observed. Our numerical results clearly explain the presence of avalanche phenomena. For tin, T and the superheating curve have been chosen to reproduce the conditions of our test, where no avalanche was found. Again, our estimate agrees with experiment but the avalanche effect may appear at slightly lower temperatures. For In, at 10% filling factor in volume, we expect avalanches to occur at T < 600 mK and even at higher T if a larger ψ is taken. This point should be seriously studied , since it may bring severe constraints on the design of a solar neutrino In detector. Also, Ga dark matter detectors will have to face similar phenomena, even at low filling factors.

5.2 HEAT CAPACITY OF A SMALL GRANULE

Here we discuss the heat capacity of a small superheated granule in the presence of a magnetic field at very low T. A nontrivial effect is the appearance of an extra contribution due to the presence of $\vec{H}_0$. The Meissner effect implies a lowering of the order parameter near the surface (Cooper pair breaking by the supercurrent) [31,32], and therefore some normal electrons are present. The total free energy of the system can be written as [33]:

$$F = F_0(H=0) + F_1(H_0) \qquad \{20\}$$

and using standard thermodynamics we get:

$$C = c_s V - T\, d/dT\, \partial F_1(H_o)/\partial T \qquad \{21\}$$

where C is the heat capacity of the granule, and V its volume. In the limit $R \gg \xi$, we can reduce the calculation of ΔF to a one dimensional problem and write [33]:

$$F_1 = \int dS\, F_{1d}(H) \qquad \{22\}$$

with $H = 3/2\, H_o \sin\theta$ and, in the limit $\kappa \ll 1$ using the Ginzburg-Landau (GL) approximation [31, 32]:

$$F_{1d}(H) = -H^2\lambda/8\pi f(0) + 2\sqrt{2}\, H_c^{\,2}\lambda/8\pi\kappa\, [2/3 - f(0) + f(0)^3/3] \quad \{23\}$$

where κ is the GL parameter, λ the penetration length and f(0) the value of the real order parameter on the surface of the sphere. For $H_o \simeq 2/3\, H_{sh}$, one has [32,2]:

$$f(0) \simeq 1/\sqrt{2}\, (1 + |\cos\theta|)^{1/2} \qquad \{24\}$$

At $t_r \ll 1$, with the approximation $\lambda \simeq \lambda_o\, (1-t_r^{\,4})^{-1/2}$, and integrating over dS, we obtain:

$$C = c_s V + c' S \qquad \{25\} \qquad \text{with:}$$

$$c'S \simeq -0.025\sqrt{2}\, R^2\, T_c^{-1}\, d/dt_r\, (H_c^{\,2}\xi) \qquad \{26\}$$

where $\xi = \lambda/\kappa$ is the GL coherence length and R the radius of the grain.

With similar approximations, and in the limit $T \ll T_c$, the required amount of energy to flip a granule at a relative threshold $\Delta H/H$ would be:

$$\Delta Q = V\, \Delta Q_V + S\, \Delta Q_S \qquad \{27\}$$

with:

$$\Delta Q_V \simeq aT_c^{\,4}/4\alpha^2\, (\Delta H/H)^2 + b\,\gamma\, T_c^{\,2}\, t_r'\int_1^\infty dz\, z^{-2}\, e^{-f z/t_r'} \quad \{28a\}$$

$t_r' \simeq \alpha^{-1/2}\, (\Delta H/H)^{1/2}$, α is given by the parametrization:

$$H_{sh} \simeq H_{sh}(0)\, (1-\alpha t_r^{\,2}), \qquad \text{and:}$$

$$\Delta Q_S \simeq 0.0125\sqrt{2}\, \pi^{-1}\, H_c(0)^2\, \xi_o\, \Delta H/H \quad \{28b\}$$

where ξ_o is the Pippard coherence length.

We have performed numerical estimates mainly for two cases:

1. Solar neutrino detection with In granules.
2. Galactic photino detection with Ga granules.

Results are given in terms of $(\Delta H/H)_D = 2\sigma$, where σ is the standard deviation of a gaussian fit to the differential superheating curve. Using this value for $(\Delta H/H)$ in {28}, about 80% of granules can be made sensitive to the ΔQ obtained. For the In detector, we require granules to be sensitive to energy deposits close to minimum ionization. We thus optimize energy resolution for electrons in the range $E \leq 1$ MeV. Granules of diameter $2\ \mu m < \phi < 4\ \mu m$ turn out to be necessary for $0.5 < \alpha < 1$ if $(\Delta H/H)_D = 0.05$, which is for the time being far from the range of any existing SSG sample. If a larger gap in $H_{sh}{}^{eff}$ is allowed, let us say, $(\Delta H/H)_D = 0.1$ (still better than any existing sample), the required diameters would be in the range $1\ \mu m < \phi < 2\ \mu m$. The operating temperature should in both cases be $T \simeq 200$ mK, but then it is not obvious that the avalanche effect can be avoided at high filling factors. Working at much higher T would degrade sensitivity and require smaller granules. The surface term {28b} turns out not to contribute significantly to the heat capacity of In granules.

In the case of gallium, the situation is different and the surface term starts to become important if one works at $T \simeq 100$ mK. With $(\Delta H/H)_D \simeq 0.05$, about $3\ \mu m$ diameter granules would be sensitive to 50 eV (approximate deposit of energy from light galactic photinos), whereas $10\ \mu m$ granules would detect 1 keV energy deposits (heavy photinos, solar axions). The coherence length of gallium has been taken to be equal to $1\ \mu m$ in the present estimates. The avalanche effect is in principle a severe problem for Ga, but dark matter experiments can be done at lower dilution coefficients since WIMP are expected to produce individual grain flips. The situation can also be improved adding impurities to the superconductive material, which is known [2, 34] to lower the value of H_{sh}.

Finally, we recall that the performance of Cd SSG detectors was estimated in Seidel's Thesis [18]. The use of Cd for recoil energy experiments deserves the same comments as for Ga, but as seen previously Cd is even more sensitive to avalanche phenomena. Such a behaviour may obviously preclude the use of Cd for double β experiments.

6. CONCLUSION AND COMMENTS

We have discussed the two important points of the physics of SSG detectors, i.e. phase transition mechanism and very low T behaviour. The phase transition mechanism determines the actual energy threshold for grain flipping, whereas very low T behaviour ($t_r \ll 1$) is expected to give the ultimate performance of the detector, assuming that no new phenomena prevent us from operating it (avalanche effect). The region $T \simeq T_c$ also provides good sensitivity [18], but signals are small as $H_{sh} \rightarrow 0$ for $t_r \rightarrow 1$.

From irradiation of large granules ($45\ \mu m < \phi < 400\ \mu m$) with α particles, we infer information on local heating phenomena. This appears to be the relevant phase transition mechanism for all the samples used. These samples have in common a high impurity content (normal state resistivity: $\rho \simeq 6 \times 10^{-8}$ to $\simeq 10^{-6}\ \Omega$.cm). For $\rho \simeq 10^{-6}\ \Omega$.cm, one may expect local heating at the μm

scale. Irradiation results with particles of lower energy (6 keV γ's) seem to confirm our analysis.

Smaller grains (10 μm < ϕ < 25 μm) have been irradiated with β^- and γ sources, and sensitivity to all of them (including 6 keV γ's) has been demonstrated. Although it is encouraging to find 8% sensitive granules for 140 keV γ's, we are far from the required efficiency and energy resolution. Work with smaller granules should improve the present figures for a detector of low energy electrons and photons.

Tests with Sn granules as low in temperature as T $\simeq$ 450 K showed no avalanche effect and the detector could be normally operated. A better knowledge of the heat transfer in the detector is necessary, but it seems possible to understand the phenomenon by simple thermodynamical considerations. It then follows that the avalanche effect may be a serious problem for most of the proposed experiments. Adding impurities makes possible to lower H_{sh} for low κ materials, as a consequence of the Gorkov-Goodman relation:

$$\kappa \simeq \kappa(\rho = 0) + 7.5 \times 10^3 \, \gamma(\mathrm{erg\ cm^{-3}\ K^{-2}})^{-1/2} \rho(\Omega\ \mathrm{cm}) \qquad \{29\}$$

This should in principle be a way to get a smaller ΔQ^L for materials such as Cd and Ga, that otherwise present large superheating.

The performance of SSG detectors at very low T depends crucially on the dispersion in H_{sh}^{eff}; i.e. the parameter $(\Delta H/H)_D$, so that there is not much to get in cooling a poor quality detector to very low T. The crucial item there is grain quality, since surface defects produce values of $(\Delta H/H)_D$ larger than 15%. Such defects may be of crystallographic origin [17], and are likely to depend on the details of the fabrication procedure. Fig. 12 shows a photograph of defects on a Billiton granule (sample c). Extramet granules often exhibit perfect hexagons, in a honeycomb configuration. Related to this phenomenon, there may be κ inhomogeneities and anisotropies [36]. Furthermore, in the presence of an applied magnetic field, low κ materials are sensitive to a thin layer of normal electrons that adds an extra term to the low t_r heat capacity of small granules.

As predicted in previous work [11, 32], large granules can produce electronic signals with high time resolution (fast risetime) provided low purity materials are

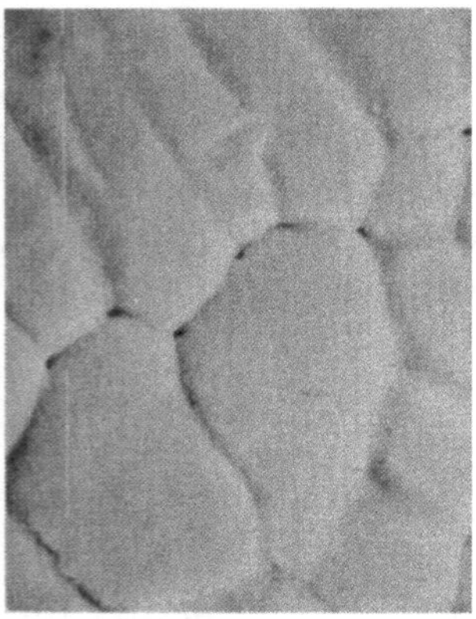

Fig. 12: Surface defects on a granule of sample c). The mark is 10 μm.

used. It is remarkable that granules in the 250 to 300 μm range presented risetimes as short as 300 nsec, and justifies the claim that a monopole experiment may become feasible.

Solar neutrino detection based on Raghavan's reaction (In granules) is likely to require very small granules, so that it is hard by now to evaluate the feasibility of the experiment. Also, avalanche phenomena are foreseen at very low T since large filling factor is used. Small three-dimensional multichannel prototypes ($\approx$ 1cm^3) with ϕ < 5 μm In granules would be an excellent workbench where to study sensitivity, energy resolution and several problems related to background. Cold Ga-As electronics appears to be a promising technique to read small In grains in real time.

Galactic photinos may be detectable with gallium granules at T $\simeq$ 100 mK, if problems related to avalanche phenonena can be handled. As for Indium, work with small prototypes seems to be a necessary step until the basic required properties (sensitivity, operativity) will actually be demonstrated.

Grain production is a crucial issue, since very small granules with good size homogeneity are required for most applications of SSG. Fig. 13 shows the size distribution of a sample of ϕ < 40 μm Sn grains produced by EXTRAMET in December 1986. The result is encouraging, but it should be proven that a similar distribution can be obtained for smaller granules.

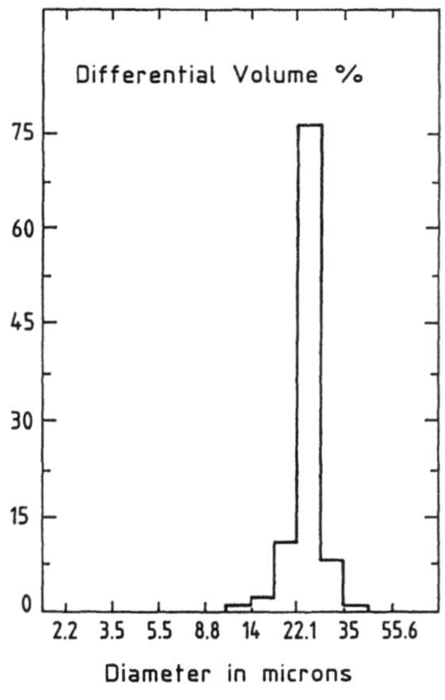

Fig. 13 : Differential size distribution of 10 μm < ϕ < 40 μm EXTRAMET granules.

Local heating may provide a way to escape some of the present limitations and get better sensitivity. It is possible that, using larger granules, larger electronic signals will be obtained without loss in sensitivity, provided the grains are made of low purity materials. We are preparing a complete study of this question. In any case, the basic properties of the detector should remain the main subject of study for the nearby period, before actual prototypes can really be built.

References

[1] See, for instance, J.P. Burger, "Superheating and supercooling in first kind superconductors" in Superconductivity, Ed. P.R. Wallace (Gordon and Breach, N.Y. 1969)

[2] For an updated introduction to the use of SSG for particle detection, see: L. Gonzalez-Mestres and D. Perret-Gallix, Preprint LAPP EXP–87–02, to appear in the Proceedings of the International School of "Astro–Particle Physics", Erice January 1987, Ed. World Scientific Pub.

[3] H. Bernas, J.P. Burger, G. Deutscher, C. Valette and S.J. Williamson, Phys. Lett. 24A, 721(1967) and C. Valette, Thesis 1971.

[4] J. Blot, Y. Pelan, J.C. Pineau, J. Rosenblatt, J. Appl. Phys. 45, 1429 (1074)

[5] See, for instance C. Valette and G. Waysand, "Détecteur supraconducteur de rayons gamma à résolution intrinsèque submillimétrique", Orsay 1976, and D. Hueber, Thesis Orsay 1980

[6] A.K. Drukier, C. Valette, G. Waysand, L.C.L. Yuan and F. Peters, Lett. al Nuovo Cimento 14 , 300 (1975). A.K. Drukier and L.C.L. Yuan, Nucl. Instrum. Methods 138 , 213 (1976).

[7] A.K. Drukier and L. Stodolsky, Phys. Rev. D30, 2295 (1984)

[8] G. Waysand, Proceedings of the Moriond Meeting on Massive Neutrinos, January 1984, Ed. J. Tran Thanh Van, p.319.

[9] R. S. Raghavan, Phys. Rev. Lett. 37. 259 (1976)

[10] L. Gonzalez-Mestres and D. Perret-Gallix, Proceedings of the Moriond Meeting on massive Neutrinos, January 1984, Ed. J. Tran Thanh Van, p. 461.

[11] L. Gonzalez-Mestres and D. Perret-Gallix, Proceedings of the Moriond Meeting on Primordial Nucleosynthesis, March 1985, Ed. Reidel, p.580. L. Gonzalez-Mestres and D. Perret-Gallix, Nuovo Cimento C9, 573 (1986), presented at Underground Physics 85.

[12] M.W. Goodman and E. Witten, Phys. Rev. D30 , 2295 (1984).

[13] A.K. Drukier, C. Freese and D. Spergel, Phys. Rev. D to appear.

[14] D. Spergel, Proceedings of the Moriond Meeting on Massive Neutrinos, January 1986, Ed. J. Tran Thanh Van, p.420.

[15] S. Dimopoulos,G. Starkman and B.W. Lynn, Phys. Lett. B 167, 145 (1986) and Mod. Phys. Lett. A1, 491 (1986).

[16] A. Pacheco, Modern Phys. Lett. 1 , 167 (1986), and these Proceedings.

[17] K. Pretzl, J. Fent, P. Freund, J. Gebauer, N.Schmitz, L. Stodolsky and G. Vestzergombi, contribution to the Moriond Meeting on New and Exotic Phenomena, January 1987, and these Proceedings.

[18] F. Von Feilitzsch, L. Oberauer and W. Seidel, Nuovo Cimento 9C, 598 (1986), presented at Underground Physics 85. W. Seidel, Thesis (1985).

[19] EXTRAMET, Zone Industrielle de Cranves–Sales, 74380 Bonne–sur–Menoge (Haute–Savoie), FRANCE.

[20] BILLITON WITMETAAL FRANCE, 48 − 58 rue Alfred Dequéant, 92000 Nanterre.
[21] T.E. Faber, Proc. Royal Soc., Ser. A, 219, 75 (1953).
[22] N. Gauthier and P. Rochon, Journal of Low Temp. Phys. 59, 225 (1985).
[23] J.E. Gueths, P.L. Garbarino, M.A. Mitchell, P.G. Klemens and C.A. Reynolds, Phys. Rev. 178, 1009 (1969).
[24] C.A. Bryant, P.H. Keesom, Phys. Rev. Lett. 4, 460 (1960) and Phys. Rev. 123, 491 (1961).
[25] D.H. Parkinson, Rep. Progr. Phys. 21, 226 (1958)
[26] N. Perrin, Preprint GPS de l'ENS, Paris 1986.
[27] L. Gonzalez-Mestres and D. Perret-Gallix, Preprint LAPP EXP − 87 − 01, to appear in the Proceedings of the Moriond Meeting on New and Exotic Phenomena, January 1987.
[28] G. Seidel, P.H. Keesom, Phys. Rev. 112, 1083 (1958)
[29] See, for instance, L.D. Landau and E.M. Lifchitz, Electrodynamique des Milieux Continus, Editions Mir, Moscou 1969.
[30] A. Baratoff, "Small superconducting samples" in superconductivity, Ed. R.D. Parks, Gordon and Breach 1969. F.W. Smith, A. Baratoff and M. Cardona, "Nonlocal effects on superheating of type I superconductors" ,in Proceedings of the XI International Conference on Low Temperature Physics, St Andrews 1968.
[31] Orsay Group on Superconductivity, Contribution to "Quantum Fluids" , presented by P.G. de Gennes , Ed. D.F. Brewer, North Holland 1966.
[32] L. Gonzalez-Mestres and D. Perret-Gallix, Preprint LAPP EXP − 84 − 05 and TH − 112 (Unpublished). L. Gonzalez-Mestres and D. Perret-Gallix, Proceedings of the EPS International Conference on High − Energy Physics, Bari July 1985.
[33] A. Baldisseri, "Comportement à très basse température d'un détecteur supraconducteur de neutrinos solaires et candidats à la matière noire galactique", Rapport de Stage DEA, Annecy March 1986.
[34] H. Parr, Phys. Rev. B14, 2849 (1976)
G. Pettersen, H. Parr, Phys. Rev. b19, 3482 (1979)
[35] B.B. Goodman, IBM J. Res. Dev. 6, 63 (1962).
[36] We are grateful to K. Pretzl and L. Stodolsky for emphasizing, in private discussions, the eventual role of κ anisotropies.

Investigation of Superconducting Tin Granules for a Low-Energy Neutrino or Dark Matter Detector

J. Fent, P. Freund, J. Gebauer, **K. Pretzl**, N. Schmitz, A. Singsaas, L. Stodolsky, and G. Vesztergombi

Max-Planck-Institut für Physik und Astrophysik, Werner-Heisenberg-Institut für Physik, Föhringer Ring 6, D-8000 München 40, Fed. Rep. of Germany

The properties of single superconducting tin granules with a diameter between 20 and 112 μm were studied between 3.26 and 1.4 Kelvin in a magnetic field. The granules were rotated around an axis perpendicular to the magnetic field axis and the superheating and supercooling fields were determined. We observed that each granule exhibits its own characteristic superheating and supercooling field which strongly depends on the rotational angle. For granules with effective superheating fields just above the critical thermodynamical field, a phase transition was observed which took place over only a part of the granule (intermediate state). Single tin granules were also irradiated with α-particles of 5.5 MeV energy. Phase transitions were clearly observed. The results are consistent with local heating.

1. Introduction

Superheated superconducting granules (SSG) have been suggested as possible detectors for solar neutrinos and dark matter particles [1,2]. The method uses the neutral current process of neutrino-nucleus elastic scattering (also for dark matter particles, if weakly interacting). The main advantages are:

a) The coherent scattering cross section is 3 orders of magnitude greater than the cross sections of other processes like, for example, inverse beta-decay. Thus, a SSG detector with a weight of a few kilograms would measure the same event rate as a multiton detector based on other processes.
b) The SSG detector responds to all kinds of neutrinos equally.

The principal difficulty with this method is, of course, the detection of a very low nuclear recoil energy E_A. Its average value is given by $\overline{E}_A = 2\,E_\nu^2/3A$ [keV] (with the neutrino energy, E_ν, measured in MeV) and comes out to be 0.9 eV (assuming Sn grains) for solar neutrinos with $E_\nu = 0.4$ MeV.

In a uniform heating model most of this energy will be transformed into heat, leading to a temperature change ΔT of the grain. This temperature jump can flip a grain from the superconducting to the normal state. This is illustrated in Fig. 1 for a tin (Sn) grain which has a superheating transition field H_{SH} and a supercooling field H_{SC}. For a grain with a radius r, a density ρ, and a specific heat c, the temperature change is $\Delta T = 3E_A/4\pi c\rho r^3$. To measure 0.4 MeV solar neutrinos, for example,

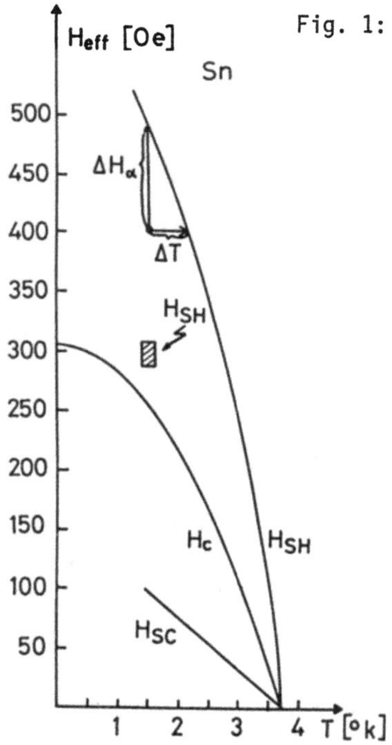

Fig. 1: Phase diagram of Sn granules with H_{SH} = superheating field, H_{SC} = supercooling field, and H_c = critical thermodynamic field

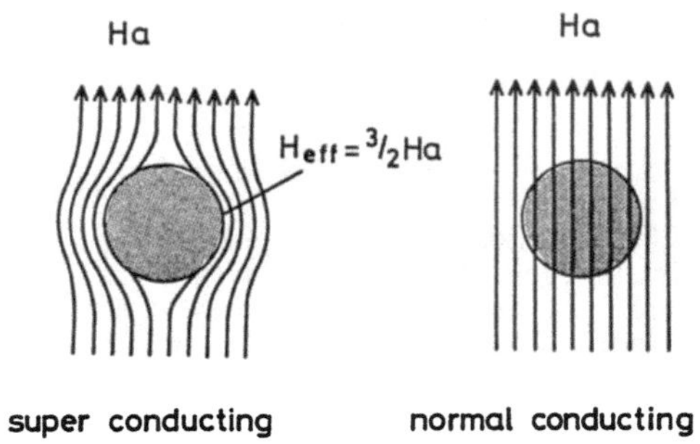

Fig. 2: Schematic view of the Meissner effect

a SSG detector would have to be made of tiny Sn grains with a diameter of 1.7 μm which are cooled down to T = 50 mK, if the minimum energy threshold would be set to 0.9 eV, corresponding to ΔT = 10 mK. The grain flip can be detected with a pick-up coil which measures the flux change due to the disappearance of the Meissner effect (Fig. 2). The expected event rates per year per kilogram SSG are 30 for solar neutrinos and 6.10^4 for weakly interacting dark matter particles with a mass of 1-2 GeV. Dark matter candidates with spin-dependent interactions (e.g. photinos) can also be detected with SSG, provided the grain material has a large nuclear spin.

Backgrounds due to natural radioactivity of the detector-material and cosmic rays are a major problem to be solved, but this will not be discussed here.

The SSG technique is still under feasibility study. Several groups [3] have reported results at this meeting. We present measurements which we performed with single Sn grains of diameter 20-112 μm. In section 2, we report on granule properties when rotated around an axis perpendicular to the external magnetic field. In section 3, we show results obtained when irradiating single grains with α-particles. Section 4 gives the conclusions.

2. Properties of Individual Sn Grains in an External Magnetic Field

This investigation was motivated by the observation [3,4] that groups of Sn or Cd grains exhibited a washed-out phase transition $\Delta H_{SH}/\overline{H_{SH}} \sim \Delta H_{SC}/\overline{H_{SC}} \sim$ 20-30% [Fig. 3]. In order to study possible surface or crystalline structure effects, we rotated indi-

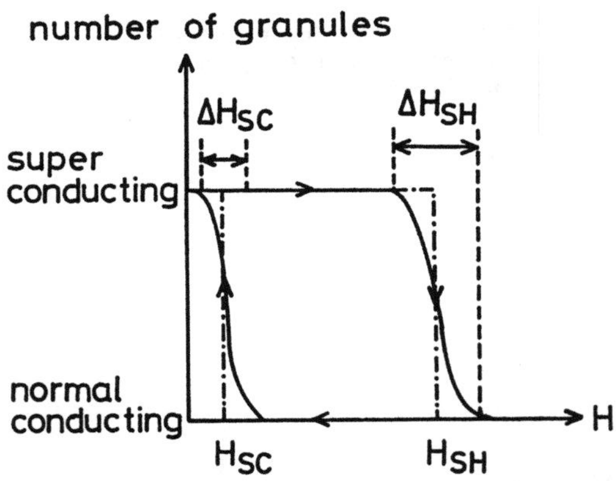

Fig. 3: Schematic view of external field hysteresis for a group of granules at fixed temperature. The experimental observation is shown as a solid line while the ideal behaviour is indicated as dash-dotted line

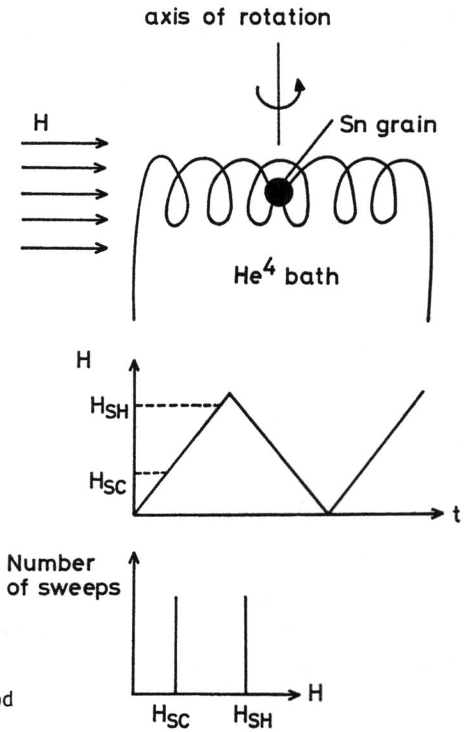

Fig. 4: Schematic view of experimental method

vidual grains around an axis perpendicular to the external magnetic field. By cycling the magnetic field at fixed temperature, we measured H_{SH} and H_{SC} as a function of the rotation angle (Fig. 4). The experiment was performed in a temperature range of 1.4 K < T < 3.26 K. The results for individual Sn grains with a diameter of 56 µm, 112 µm, and 20 µm are shown in Fig. 5, 6, and 7 respectively for various temperatures (see also ref. 5). The measured variations $\Delta H_{SH}/\overline{H_{SH}}$ and $\Delta H_{SC}/\overline{H_{SC}}$ of individual grains when rotated in an external field came out to be as large as 30%, thus explaining the magnitude of the phase transition smearing observed for a group of grains.

For granules with effective H_{SH} (in Fig. 1 shown as H'_{SH}) very close to the critical thermodynamical field H_c, we measured grain flip signals which were very much smaller than expected for a phase transition of an entire grain. This observation is consistent with what one would expect if the granule were subdivided into superconducting and normal zones (intermediate state). In some cases, the superconductivity of the grain is only partially broken, up to the point where $H_{eff} < H_c$ (with H_{eff} being the effective field at the equator of the grain where the field lines are compressed, Fig. 2).

Fig. 5: Superheating and supercooling field for a Sn grain with 56 µm diameter at various temperatures

Fig. 6: Superheating and supercooling field for a Sn grain with 112 µm diameter at various temperatures

Fig.5

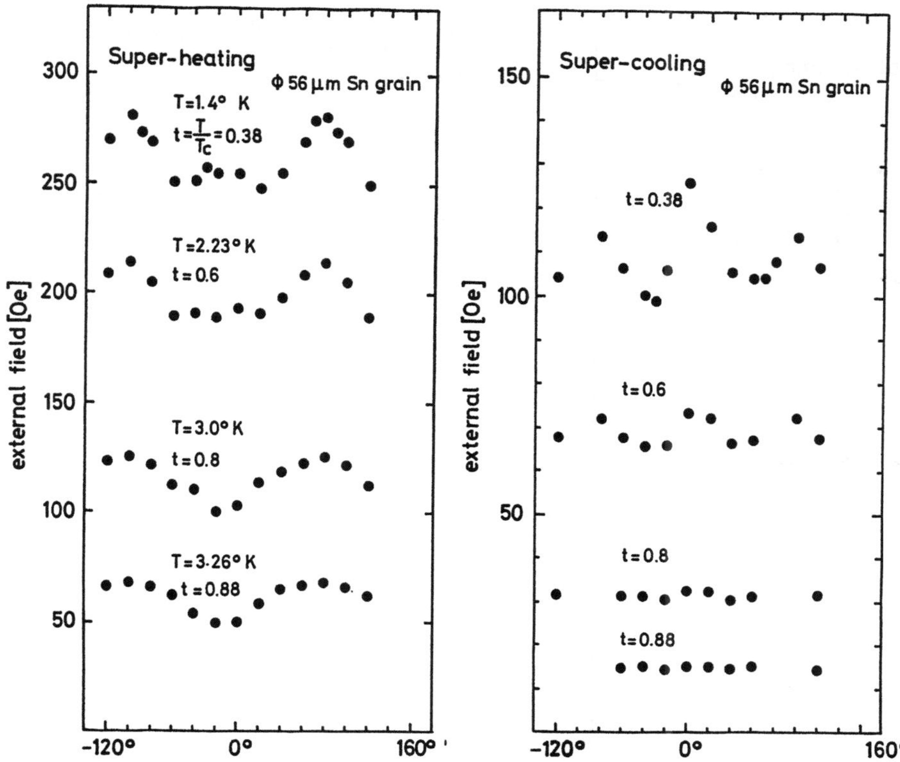
Super-heating
Φ 56 μm Sn grain
T=1.4° K
t=T/Tc=0.38
T=2.23° K
t=0.6
T=3.0° K
t=0.8
T=3.26° K
t=0.88
external field [Oe]
300
250
200
150
100
50
-120°
0°
160°
Super-cooling
Φ 56 μm Sn grain
t=0.38
t=0.6
t=0.8
t=0.88
external field [Oe]
150
100
50
-120°
0°
160°

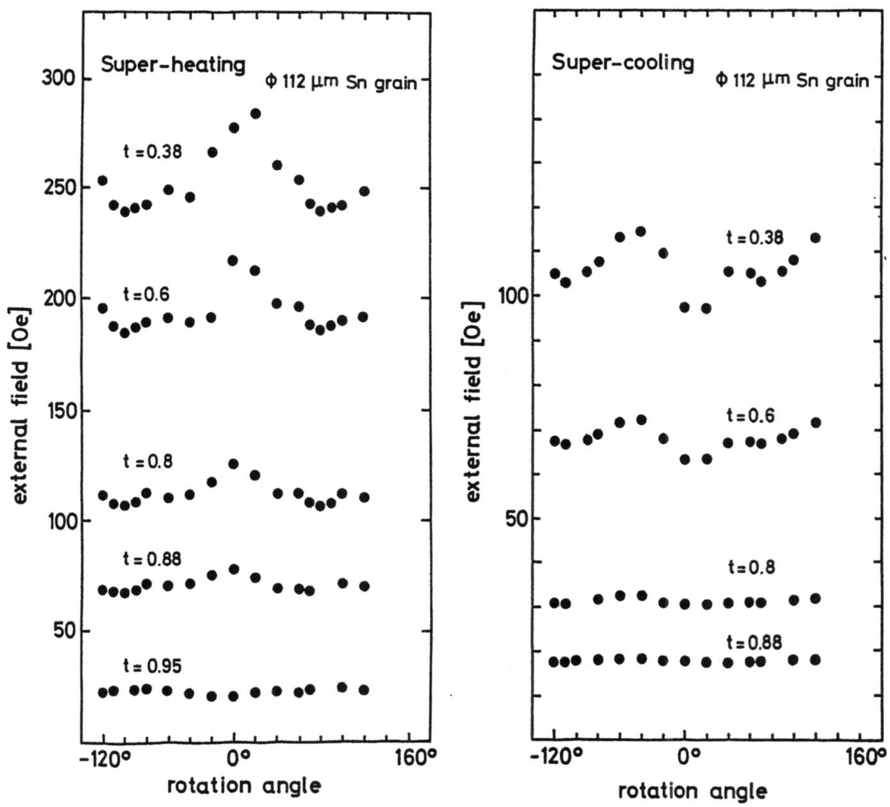
Super-heating
Φ 112 μm Sn grain
t=0.38
t=0.6
t=0.8
t=0.88
t=0.95
external field [Oe]
300
250
200
150
100
50
-120°
0°
160°
rotation angle
Super-cooling
Φ 112 μm Sn grain
t=0.38
t=0.6
t=0.8
t=0.88
external field [Oe]
100
50
-120°
0°
160°
rotation angle

Fig.6

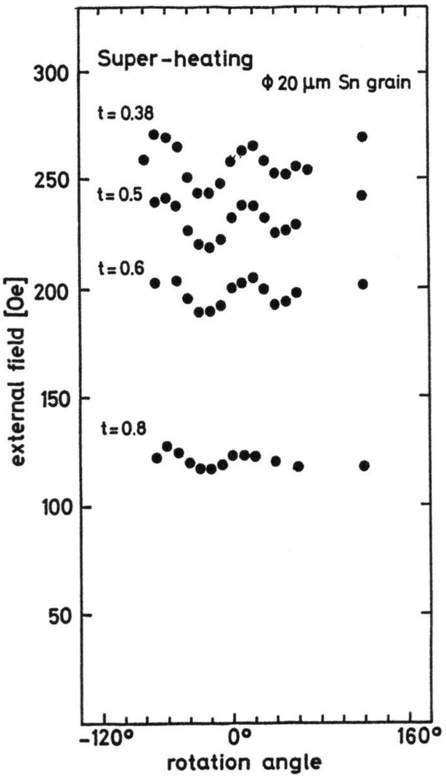

Fig. 7: Superheating field for a Sn grain with 20 μm Sn grain at various temperatures

3. Irradiation of Individual Sn Grains with α-particles

We have chosen an α-source (Am^{241} source with $E_\alpha \sim 5.5$ MeV) for the irradiation experiment since α-particles, in central collisions, lose all their energy in the grain. Their penetration length in Sn is 14 μm. The α-source, with an activity of 6.35 μCi, was mounted at a distance of about 200 μm from the grain inside the pick-up coil. In order to reactivate the grain after it flipped into the normal state, the external magnetic field was cycled as shown in Fig. 4. The helium bath was kept at T = 1.4 K. Due to the energy release of the α-particle in the grain, the magnetic field at which the flip of the grain occurs is ΔH_α below H_{SH} at T = 1.4 K (Fig. 1). In the case of global-heating, ΔH_α is proportional to the energy release of the α-particle and inversely proportional to r^3 for small ΔT. Fig. 8a, b, and c show the measured distributions of the magnetic field at which the phase transition occurred for individual Sn grains with 56 μm ϕ, 100 μm ϕ, and 30 μm ϕ, respectively. As shown schematically in Fig. 8, the measurement was performed at various irradiation angles. For monoenergetic α's, a sharp peak at H_{SH}-ΔH_α would be expected. However, because of the finite energy loss of α's in the helium bath and non-central collisions with the grain, this distribution is broadened. The change of the center of gravity of the distributions at 80° and 70° indicates that the grains are more sensitive if α-parti-

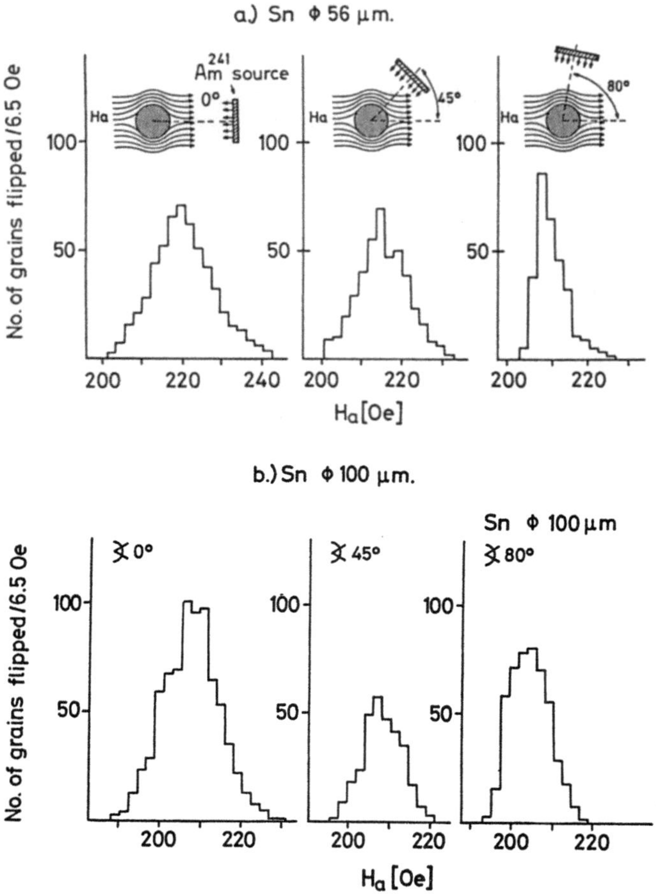

Fig. 8 a, b: The phase transition field H_a of a single Sn grain irradiated by α-particles at different angles is shown for $T = 1.4$ K for a Sn grain a) with 56 μm diameter, and b) with 100 μm diameter

cles hit at the equator of the grains, where H_{eff} is largest, than at the pole. The activity of the α-source used for the irradiation of the 56 μm and 100 μm ϕ grains was strong enough that each grain flip was induced by α-particles. For the 30 μm ϕ grain, a new and somewhat weaker α-source was employed. In this case, the observed grain flips are sometimes due to α-particles (broad distributions in Fig. 8c) and sometimes due to the external magnetic field sweep above H_{SH} (pronounced H_{SH} peaks at the right side of the histograms in Fig. 8c). The maximal ΔH_α came out to be 41(41)Oe at 0°, 33(25)Oe at 45°, and 24(25)Oe at 80° rotation angle for 56 μm(100μm) ϕ Sn grain and 62 Oe at 0°, 68 Oe at 45°, and 59 Oe at 70° for 30 μm ϕ Sn grain. The proportionality $\Delta H_\alpha \sim 1/r^3$, expected for global heating, is not observed. Both observations are consistent with local heating, where the superconductivity of the grain starts to be broken locally around a region where the α-particle releases most of its energy. In contrast to the nucleus recoil in coherent neutrino scattering, the

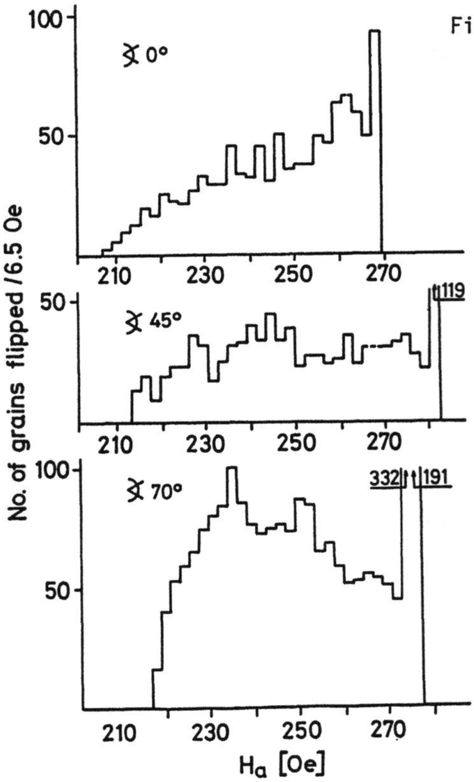

Fig. 8c: The phase transition field H_a of a single Sn grain irradiated by α-particles at different angles is shown for T = 1.4 K for a Sn grain with 30 µm diameter

c.) Sn ϕ 30 µm

energy loss of α's is mostly concentrated on the surface of the grains and well above the ionisation energy. This leads to a fast local break-up of cooper pairs, which apparently occurs before the grain is globally heated by phonons.

4. Conclusions

In a feasibility study of a SSG detector for low-energy neutrinos and dark matter, we performed measurements with individual Sn grains of 20-112 µm in diameter. It was shown that the smearing of the supercooling and superheating transition fields, previously observed with a group of grains, is largely due to the different behaviour of individual grains. When irradiating Sn grains with α-particles, local heating of the grains seems to be the dominant process.

References

1. A. Drukier and L. Stodolsky: Phys. Rev. D30, 2295 (1984)
2. M.W. Goodman and E. Witten: Phys. Rev. D31, 3059 (1985)
3. see L. Gonzales-Mestres, D. Perret-Gallix, A. Kotlicki, G. Waysand, A. Bellefon, F. v. Feilitzsch, F. Probst, W. Seidel these proceedings
4. F. v. Feilitzsch et al.: Il Nuovo Cimento, Vol. 9C, N2 598 (1986)
5. J. Feder and D.S. McLachlan: Phys. Rev. 177, 763 (1969)

SQUID Detection of Superheated Granules

M. Le Gros[1], B.G. Turrell[1], M.J.C. Crooks[1], **A. Kotlicki**[2], *and A.K. Drukier*[3]

[1]Department of Physics, University of British Columbia, Vancouver, B.C., Canada, V6T A6
[2]Institute of Experimental Physics, Warsaw University, Hoza 69, PL-00-681 Warsaw, Poland
[3]Harvard-Smithsonian Center for Astrophysics, 80 Garden St., Cambridge, MA 02138, USA

1. Abstract

We report two tests designed to study the practicality of a superheated superconducting colloid detector using a SQUID readout system. In the first test, the individual 'flips' of ten 15μm radius tin grains were observed as the temperature was swept through the superheated superconducting — normal phase transition. In the second test, we were able to observe transitions induced by 90 keV γ-rays in a colloid of 5 μm radius grains in epoxy.

2. Introduction

The last few years have seen an increasing interest in the development of low-temperature particle detectors, which include the superheated superconducting colloid detector (SSCD) [1–7], the superconducting tunnel junction [8], the crystal bolometer [9–11] and the hybrid detector [12]. Our own interest is in the SSCD, and in this paper we will discuss recent investigations performed at the University of British Columbia. Although this detector was developed for applications in high-energy physics and photon detection, it is also a promising candidate for detecting neutrinos [13] and weakly interacting massive particles (WIMPS) [14–16]. The ideal SSCD consists of a large number of identical micron-sized grains immersed in an appropriate dielectric and prepared in a superheated superconducting state.

The deposition of a few keV of energy into a given grain will then 'flip' it into the normal state, thereby producing a change of magnetic flux which can be detected in a pick-up coil. The energy needed to flip the grain depends on the grain material, size and operating temperature.

Two tests were performed. In Test 1, the flipping of individual 15 μm radius grains was detected as the temperature was swept, while in Test 2, the effect of low-energy γ-radiation on a colloid sample of 5 μm radius grains was investigated. The experiments were performed in a standard $1K$ pumped ^4He cryostat. A superconducting solenoid supplied an external field B on the sample, and both this field and the temperature of the sample were computer-controlled, allowing the sample to be taken through various paths in the B–T phase diagram. The changes in flux produced by grain flips were detected by a SQUID connected to the pick-up coil, which, together with the $1K$ cryostat and the solenoid, were shielded in a superconducting lead jacket. The signal is proportional to the magnetic field and the grain volume, and depends on the geometry of the pick-up coil and the various inductances in the detecting system.

3. Experimental Tests

In this section, we describe the two tests. The samples used in these tests were selected from batches prepared by ultrasound disintegration of 99.99% pure tin. Size selection was performed by filtering the grains through wire meshes.

3.1. Test 1. Single–Grain Signal

In this experiment, a line of 15 μm radius grains was mounted on a copper strip, one end of which was attached to a regulated heat source, while the other end was heat–sunk to the $1K$ pot via a large thermal resistance. With this arrangement, the temperature could be swept with each of the ten grains being at a slightly different temperature. The ten grains were selected using a microscope so that good spheres of the same size were used. Each grain was mounted on a piece of teflon tape which was then placed on copper cold finger so that there was no electrical contact between the grain and the copper. Good thermal contact between grain, tape and copper was effected by using high thermal conductivity grease which also encapsulated the grain.

Signals were detected using a commercial SHE System 330 RF SQUID mounted in the $4.2K$ bath. The output voltage of the SQUID was digitized using a multichannel analyser (Tracor 1710) with 12–bit digitizer, controlled by the computer. As well as recording the output voltage, the SQUID re–sets were recorded using a home–made flux counter.

Figure 1 shows the signals obtained from the ten 15μm radius grains as the temperature was swept in the cycle $ABCDA$ in the phase diagram Fig. 2 [17], where A is below the supercooling transition point D, and C is above the superheating transition point B. The small temperature dependence of the SQUID signal in the A and C regions was due to paramagnetic impurities in the copper cold finger and the noise came from vibrations transmitted through the cryostat mounts. These were subsequently reduced to a level at which 5 μm radius grain signals could be observed with a signal/noise ratio of 10.

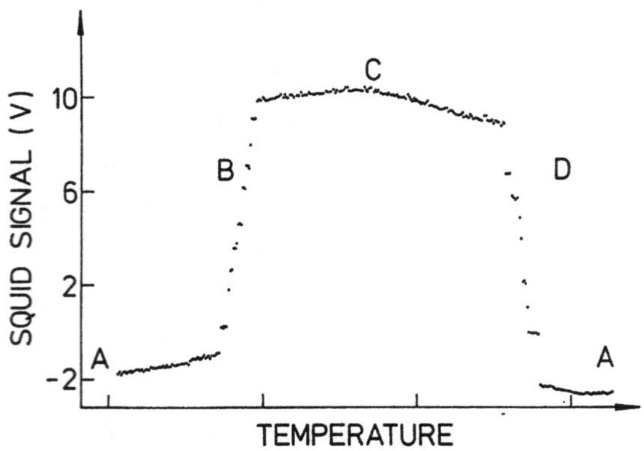

Fig. 1. The sequential phase transitions of ten 15 μm radius grains. The points $ABCD$ are shown on the B–T phase diagram, Fig. 2.

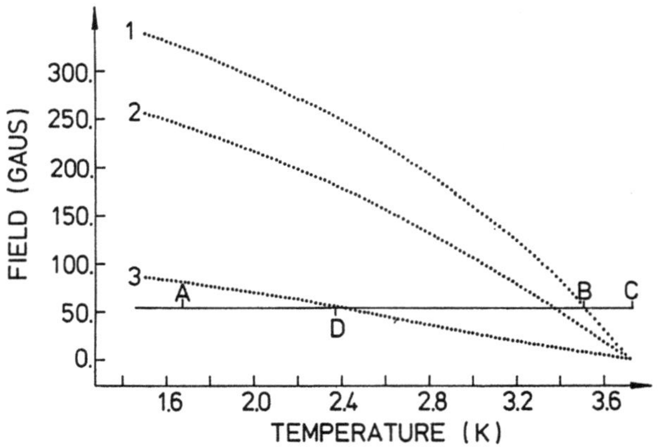

Fig. 2. The B–T phase diagram of the grains [17]. Boundary 1 represents the superheated transition line, boundary 2 is the thermodynamic transition line and boundary 3 is the supercooling transition line.

3.2. Test 2. Radiation Test

For this experiment, a sample was prepared consisting of about $10^4 \mu m$ radius grains embedded in an epoxy resin (Emerson and Cummings Stycast 1266). This sample was loaded into a copper capillary connected directly to a temperature–regulated copper block. The source of γ–radiation was a ^{67}Ga sample placed inside the cryostat, and a 1 cm thick copper block could be moved by magnetic remote control to absorb the γ–rays. When the block was in place the 90 keV γ–rays were strongly attenuated (to about 2% of the unblocked radiation), while the higher energy γ–rays suffered only small attenuation. However, the low–energy γ–rays have by far the highest probability of flipping a grain. The experimental set–up is shown in Fig. 3.

It should be noted that no grain selection was made other than the initial filtering by the meshes. Thus there were significant variations in both the size and shape of the particles. These variations, coupled with inhomogeneity of the grain distribution within the epoxy, which cause a spread of dipolar fields, produce a relatively large spread in the individual grain transition temperatures, as can be seen in Fig. 4. The effective colloid transition width increases with field. A typical radiation test comprised the following sequence of operations:

(1) The colloid sample was collected in $B = 0$ to $T < 1.9K$ ensuring that all the grains were superconducting.

(2) The external field was then increased to the required value.

(3) The temperature was then swept slowly upward until the SQUID signal indicated the beginning of the superheated phase transition.

(4) The temperature was stabilized, and, after the SQUID output indicated that the sample had equilibrated and that thermally–induced transitions had ceased, the γ–ray aperture was alternately opened and closed while the SQUID output was recorded.

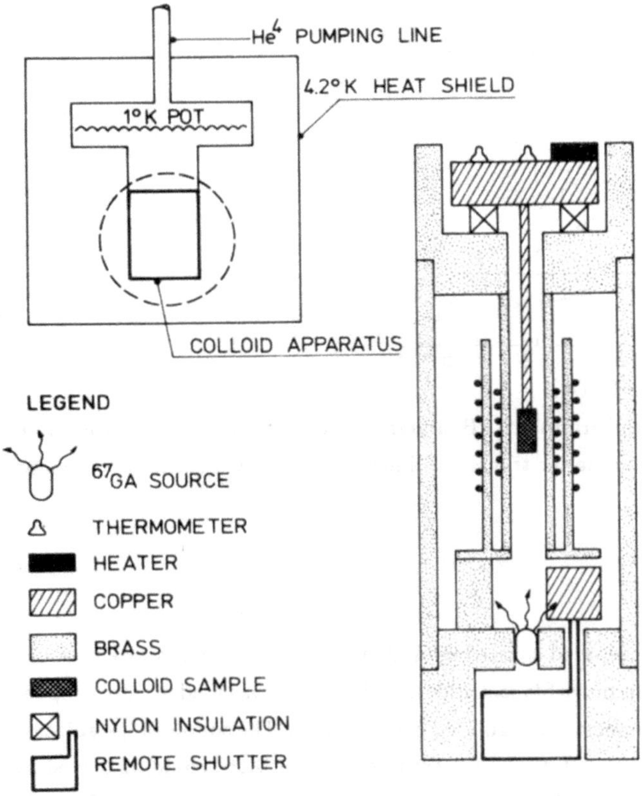

Fig. 3. The design of the apparatus for Test 2. The coils around the colloid sample are the pick–up coil and the superconducting solenoid.

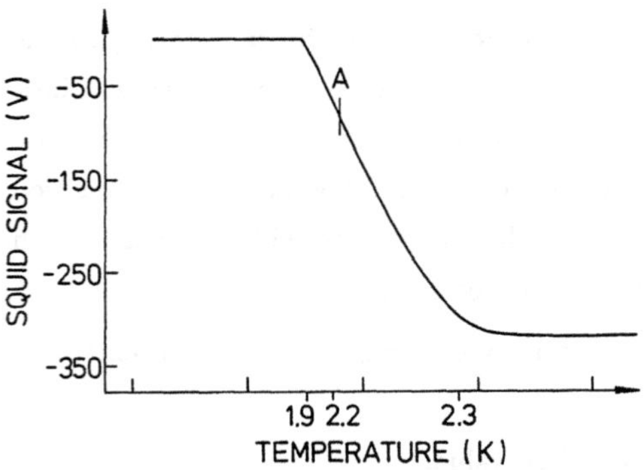

Fig. 4. The deconvoluted SQUID signal for the colloid at $B = 0.031T$. Point A is a typical temperature setting for the radiation Test 2.

Note that in the 'opening' and 'closing' operations in Step 4 the temperature had first to be lowered to off-set heating effects and then re-set. The shutter magnet was activated only in these operations and was off during the measurements. Also the SQUID was deactivated by a heat-switch on the pick-up circuit during opening and closing and whenever the field was changed, e.g. in Step 2.

The results of a radiation test at $B = 0.031T$ and $T = 2.2K$ are shown in Fig. 5 and Fig. 6 which show slow and fast scans respectively. In Fig. 5B, the aperture is closed and the signal (change in voltage) is due to low-probability transitions induced by high-energy γ-rays which pass through the shutter. In Fig. 5A, the aperture is open. For the fast scan with open aperture shown in Fig. 6A the individual grain flips can be seen.

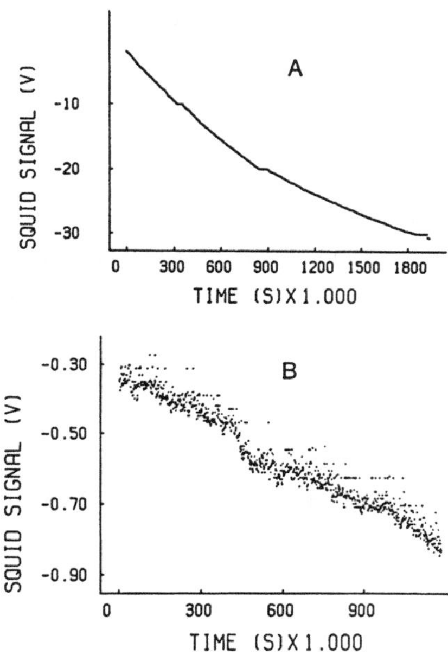

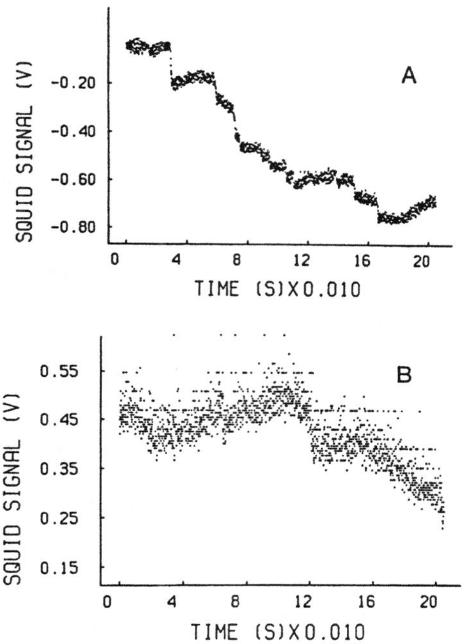

Fig. 5. The SQUID signal from the superheated colloid with the γ-ray shutter open (A) and closed (B). The field $B = 0.031T$ and temperature $T = 2.2K$.

Fig. 6. A faster scan showing the SQUID grain $\times$ 100. In (A) the shutter is open; in (B) it is closed. Note the single grain flips in (A).

While observing grain flips the temperature was reduced until the flipping ceased. By this procedure we estimated the maximum temperature gain caused by a γ-ray hitting a grain to be approximately 50 mK. Thus the only grains that can be driven normally are those with their transition temperatures within 50 mK of the quiescent temperature. This means that the effective grain filling factor f_e is much less than the mass filling f_m. For example, for $B = 0.031T$

we estimate $f_e/f_m = 0.05$. Taking this into account, the fraction of γ–rays absorbed and the solid angle presented by the colloid sample, we define a quantum detection efficiency (QDE) as the ratio of the grains that can and do flip when hit by a γ–ray to the number that can flip. We then estimate the QDE $\approx$ 85%.

4. Conclusion

We have built a practical SQUID read–out system that is sensitive enough to detect individual 5 μm radius grain transitions. Also we have demonstrated that a stable superheated colloid state can be prepared which becomes unstable under irradiation of 90 keV γ–rays. For this SSCD we estimate the QDE $\approx$ 85%.

This is very encouraging to the prospect of building a SSCD to detect neutrinos and dark matter. The greatest problem to overcome is the production of a homogeneous colloid of grains that are defectless and have uniform size and shape.

5. Acknowledgments

This work was financially supported by the Smithsonian Astrophysical Observatory. We wish to thank Dr. M. Hasinoff and Dr. A. Olin for helpful discussions and providing the ^{67}Ga source, and the following U.B.C. summer and engineering students who, at various times, assisted on the project: D. Macquistan, E. Zarmes, N. Wright, B. Currel and J. Forrest.

6. References

[1] H. Bernas, J.P. Burger, G. Deutscher, C. Valette, and S.J. Williamson: Phys. Lett. **24A**, 721 (1967)

[2] A.K. Drukier and C. Valette: Nucl. Instr. and Meth. **105**, 285 (1972)

[3] A.K. Drukier, C. Valette, and G. Waysand: Lett. Nuovo Cimento **14**, 300 (1975)

[4] A.K. Drukier and L.C.L. Yuan: Nucl. Instr. and Meth. **138**, 213 (1976)

[5] D. Huber, C. Valette, and G. Waysand: Nucl. Instr. and Meth. **165**, 201 (1979)

[6] A.K. Drukier: Nucl. Instr. and Meth. **173**, 259 (1980)

[7] A.K. Drukier: Nucl. Instr. and Meth. **201**, 77 (1982)

[8] N. Coron, G. Dambier, G.J. Focker, P.G. Hansen, G. Jegoudez, B. Jonson, J. Leblanc, J.P. Moalic, H.L. Raun, H.H. Stroke, and O. Testar: Nature **314**, 75 (1985)

[9] E. Fiorini and T.O. Niinikoski: Nucl. Instr. and Meth. **224**, 83 (1984)

[10] S.H. Moseley and J.C. Mather: J. Appl. Phys. **56**, 1257 (1984)

[11] D. McCammon, S.H. Moseley, J.C. Mather, and R.F. Mushotzky: J. Appl. Phys. **56**, 1263 (1984)

[12] B. Cabrera, L.M. Krauss, and F. Wilczek: Phys. Rev. Lett. **55**, 25 (1985)

[13] A.K. Drukier and L. Stodolsky: Phys. Rev. **D30**, 2295 (1984)

[14] M.W. Goodman and E. Witten: Phys. Rev. **D31**, 3059 (1985)

[15] I. Wasserman: Phys. Rev. **D33**, 2071 (1986)y

[16] A.K. Drukier, K. Freese, D.N. Spergel: Phys. Rev. **D33**, 3495 (1986)

[17] J. Feder and D.S. McLachlan: Phys. Rev. **177**, 763 (1969)

VLSI Superconducting Particle Detectors

O. Liengme†

Materials Sciences Division, Argonne National Laboratory,
Argonne, IL 60439, USA

1 Introduction

Superconductors exhibit many attractive properties for particle detection. Among these, spin sensitivity should be mentioned, as well as low threshold energy (as set by the superconducting energy gap or the critical temperature) and potentially high signal to noise ratio. The purpose of this paper is not to review these numerous applications, but rather to present the hotspot model and define its validity range. This concept leads to a class of superconducting detectors. Predictions on particle-induced switching of Josephson junctions and superconducting strips or wires are obtained from this hotspot model. These results agree well with experimental data from the literature. Finally, the propagating hotspot is suggested as a method for very high resolution particle position detection and imaging.

2 The Hotspot Approximation

A superconductor submitted to an impinging particle experiences a strong nonequilibrium state until relaxation is achieved. (See Reference [1] for a general description of nonequilibrium superconductivity). The energy spectrum of a superconductor is shown in Fig. 1, where the zero of kinetic energy has been defined at the Fermi energy E_F, thus reversing the sign of the kinetic energy of holes with respect to that of electrons. This spectrum is characterized by the energy gap Δ at the Fermi energy, the energy of the k electronic state being :

$$E_k = (\Delta^2 + \xi_k^2)^{1/2} \qquad (1)$$

where ξ_k measures the departure from the Fermi energy $\xi_k = \epsilon_k - E_F$.

We assume that an energy exchange $\delta E \gg \Delta$ occurs during the particle interaction with the superconductor. The first excited electrons will decay relatively fast as scaled by the inelastic scattering time τ_E (typ. 10^{-9} to 10^{-12} [s]) by emitting 2 Δ phonons that will break further Cooper pairs.

After this initial relaxation and for $E \approx \Delta$, recombination occurs over the slower time τ_Δ (see Fig. 1) :

---

†Present Address : 1253 Vandoeuvres, Genève, Switzerland.

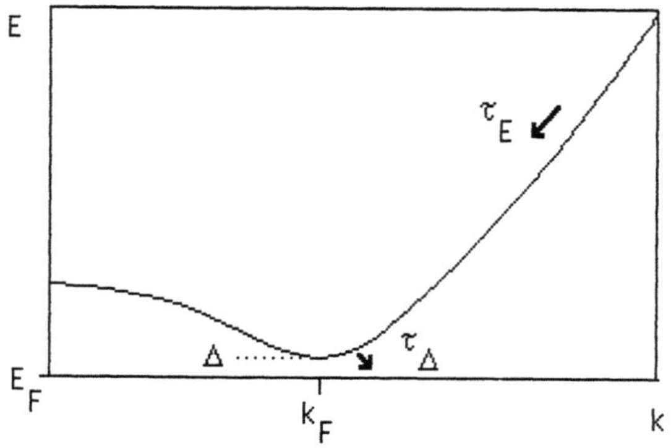

Fig. 1: *Excitation spectrum of a superconductor. The* Δ *structure near* T_c *has been magnified, since* $\Delta \ll E_F$. *Inhomogeneous excitations injected at* $E \gg \Delta$ *relax first with the inelastic scattering time* τ_E. *For* $E \sim \Delta$, the *quasiparticle recombination time is* $\tau_\Delta > \tau_E$.

$$\tau_\Delta \approx (k_B T^* / \Delta)\, \tau_E \qquad (2)$$

where T^* is the local effective temperature of the electrons.

Thin barrier junctions have been used [2,3] in a charge counting mode to exploit the slower τ_Δ for proportional counters, the quasiparticle number after relaxation at the gap edge being proportional to the energy released by the particle. Best resolution for 6 keV soft X-rays is better than 100 eV [4].

It should be noticed that the Cooper pairs rearrange quasi-instantaneously as compared to the hotspot growth, since the superfluid kinetic time [5] is :

$$\tau_s = m\sigma/n_s e^2 \approx 0.5\ 10^{-12}\,[s]/(T^*-T_b) \ll \tau_E \qquad (3)$$

at low temperatures, where T is expressed in [K].

In the hotspot approximation [6], we will consider the effective temperature T^* at (and above) T_c of high-energy electrons, i.e. before recombination. This effective temperature has a nearly gaussian profile for $T^* > T_c \gg T_b$, that is expanding with time as sketched in Fig. 2. The normal state hotspot size grows by diffusion with the characteristic time τ_E , until collapse occurs when $T^* < T_c$ and quasiparticles recombine within τ_Δ.

The maximum hotspot size D may be estimated using $D \approx 2\sigma$ at $2T_c = T^*_{max}$, using the specific heat and thermal conductivity of the material, and conveniently scaled [7] by the energy deposited by the particle. For most Nb-based superconductors and for soft X-rays of energy 6 keV it is estimated that $D \approx 0.2$ µm. For alpha particles $D \approx 4$ µm/MeV and per micron penetration depth is found experimentally, which is in good agreement with the estimated value.

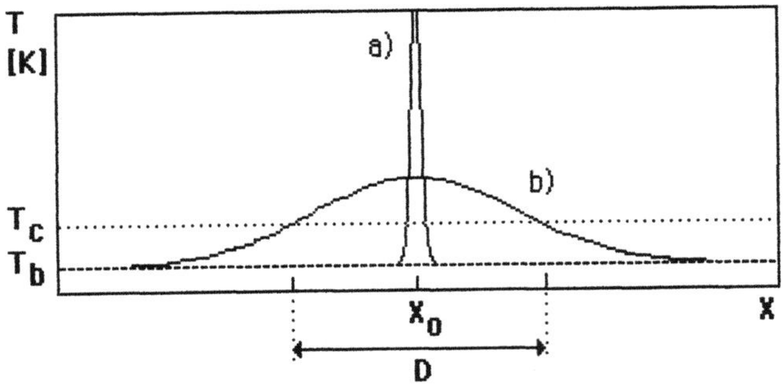

Fig. 2: *Cut view of the effective temperature profile as a function of position as induced by a particle impact at X_0 at two different times, a) shortly after the particle event and b) at the maximum hotspot size* D.

3. Particle-Induced Switching
3a) Hysteretic Josephson Junctions

In general, the effects of fluctuations are difficult to handle, resulting in a poor estimation of efficiency and signal to noise ratio. However, the case of a Josephson Junction is of particular interest, its dynamics being very well known. A particle-induced hotspot can be used to switch a hysteretic junction from zero voltage to the dissipative state [7]. Such a junction of area A, submitted to a current bias I, is illustrated in Fig. 3a with a hotspot of area $A_p = \pi D^2/4$. In order for the junction to switch into the dissipative state (Fig. 3b), the hotspot size D (and/or the current I) has to be large enough compared to the junction area A (resp. the critical current I_c).

At current threshold I_t, the critical current density $J_c = I_c/A$ will be reached in the remaining part of the area $A-A_p$, switching the junction. Therefore, we expect the corresponding current threshold I_t [7,8]:

$$I_t = I_c \, [1 - A_p/A]. \qquad (4)$$

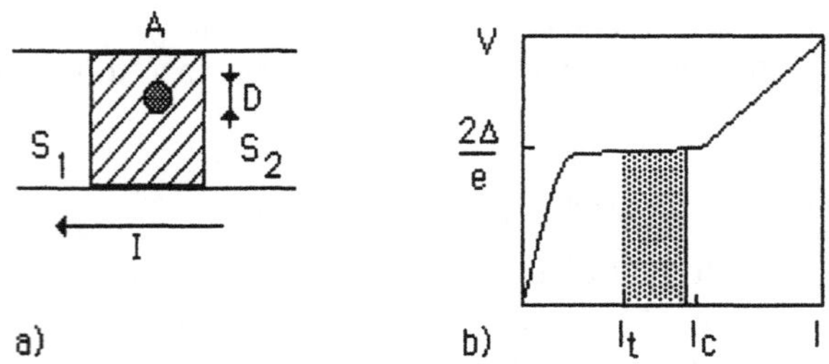

Fig. 3: *a) Josephson Junction of area* A *of electrodes* S1 *and* S2, *with a hotspot of diameter* D *and submitted to a current* I. *b) I-V characteristics of the junction of critical current* I_c *and predicted current threshold* I_t.

For the sake of simplicity, it will be assumed below that monoenergetic interacting particles lead to a single hotspot size D, regardless of phonon and/or electron escape.

We now address the question of the signal to noise ratio of such a detector. Very close to I_c, the junction may also switch because of thermal fluctuations at T > 0 [K]. The lifetime τ_J of a large hysteresis (low damping) junction with respect to this thermal activation is calculated within the RSJ model [9]:

$$\tau_J = [\omega/2\pi) \cdot \exp(-\gamma\epsilon/2) \qquad (5)$$

where ω is the attempt frequency of the junction, γ is the dimensionless temperature parameter $\gamma = \hbar I_c/ek_BT$ and ϵ is a barrier height.

On the other hand, we estimate the particle activation lifetime τ_p from eqn. 4 and assuming a unique hotspot area A_p :

$$\tau_p = \varphi^{-1} \cdot \Theta(I-I_t), \qquad (6)$$

where φ is the particle event rate $\varphi = \Phi \cdot A$, Φ being the particle flux and Θ the Heaviside function.

Statistically, switching due to fluctuations can be dissociated from those due to particle events in an experiment where the junction is biased with a sawtooth current source of constant rate of change dI/dt. Then, the total switching probability of the junction is easily calculated [10]:

$$P(I) = (\tau_{eff}(I) \cdot dI/dt)^{-1} \left[1 - \int_0^I P(u)du \right] \qquad (7)$$

where the term in brackets represents the probability of finding the junction unswitched at the current I. In our case of a particle-irradiated junction the total switching rate should read:

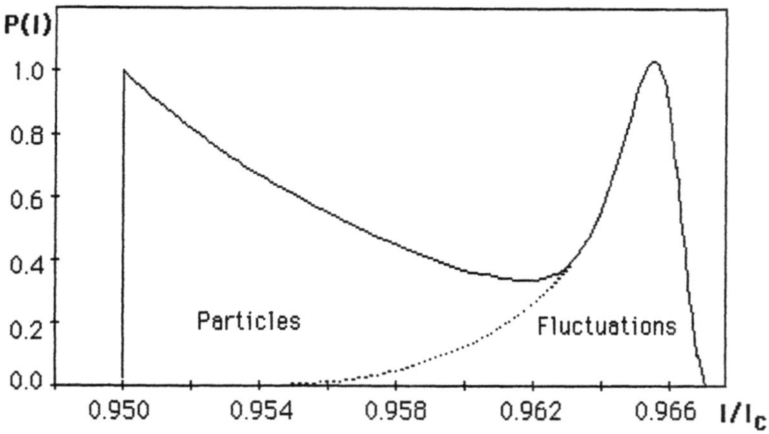

Fig. 4: *Calculated switching probability* P(I) *as a function of the reduced current bias* I/I_c. *The dotted line separating particle-induced switching (left) from fluctuations (right) is only indicative. Notice the sharp* P(I) *threshold at* $I_t/I_c = (1-A_p/A) = 0.95$ *corresponding to* $A/A_p = 20$.

$$\tau_{eff}^{-1} = \tau_p^{-1} + \tau_J^{-1}. \tag{8}$$

Figure 4 illustrates this complex P(I) as a function of the reduced current I/I_c, where the effect of thermal fluctuations at 1 K are taken into account. Junction parameters are I_c= 100 µA, C = 10 pF, A = 20 A_p, $\Phi \cdot A = 10^{-3}\ s^{-1}$, and the sweep rate dI/dt is matched to about 2 particles per thermal fluctuation. Since the attempt frequency ω is proportional to the Josephson plasma frequency $\omega_J = [\hbar I_c/2eC]^{1/2}$, changing the junction dimensions does not affect τ_J, for a given barrier thickness. Preliminary lifetime results on junctions irradiated by alpha particles [7] are in qualitative agreement with equation 8.

Hotspot detection of 6 keV soft X-ray would require estimated small-area junctions A ≤ 0.8 µm^2, assuming an area ratio A/A_p = 25. In any case, a series of junctions can be used for larger detector areas, but the low gap voltage prevents unfolding the switched junction position with semiconducting diodes, unless perhaps for $T_c \geq 100$ K as recently observed [11].

3b) Superconducting Wires

Superconducting wires can be used for larger hotspot detector areas. Alpha particle switching experiments have been reported [12,13], showing current threshold and asymmetric voltage spikes. A phenomenological hotspot model was used by the authors to discuss quantitatively such experiments in various current bias regimes.

Figure 5 shows a superconducting strip of width w, submitted to a current bias $I_b < I_c$. If the wire thickness t is small as compared to the hotspot size D, as well as if the particle leaves a track of excitation, the 2-D current threshold I_t will be :

$$I_t/I_c = 1 - D/w, \tag{9}$$

corresponding to the critical current density being reached in the w-D region. Unfortunately, the effect of fluctuations is not calculable in the general case of wires, since the wire nonequilibrium dynamics should depend on the dimensionality of the superfluid during the hotspot growth.

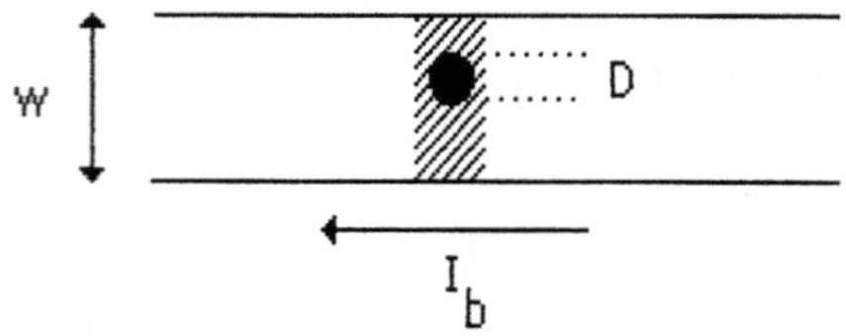

Fig. 5 : *A superconducting wire of width* w *biased at a current* I_b *submitted to a particle hotspot of width* D, *exhibits hotspot amplification (hatched area).*

The hotspot size $D \approx 22$ µm can be estimated for 5.3 MeV alphas [13], a result that agrees with that of junctions provided one takes into account the energy released by the particle as a scaling parameter.

Above threshold, two situations may occur, according to the balance of thermal energies [13,14,15]. On the one hand, the Joule energy density is :

$$W_J = \rho_N J^2 \quad (10)$$

where J is the bias current density and ρ_N is the normal state resistivity.

On the other hand, this dissipated energy is balanced by the thermal energy extracted by the medium :

$$W_M = [C_v^2/K] (T^* - T_b) \quad (11)$$

where C_v is the specific heat and K is the thermal conductivity in the hotspot.

Obviously, the hotspot will collapse when $W_J < W_M$. Particle interaction events are observed [12,13] as resistance spikes with a rise time of about $10 \cdot \tau_E$ and the fall time determined by this energy balance. It is a typical bolometer application with a (small) hotspot amplification and current threshold. Notice that an external device (e.g. an X-Y cross-line array with correlator) is required to determine the particle position.

However, the case of $W_J > W_M$ leads to a propagating hotspot [14,16] and deserves special attention, since a propagating hotspot triggered by a particle (such as observed for alpha particles [15]) may lead to a very high resolution position reading detector (see below).

For $W_J > W_M$ the zero dissipation state of the entire wire may be regarded as metastable. The amplified hotspot propagates with a speed v_H because of Joule over-dissipation [14] :

$$v_H^2 = W_J / W_M \quad (12)$$

for large I_b where $W_J \gg W_M$. If metallic thermal conduction (e.g. along the wire) dominates [16], (as well as in the case of a presumed temperature mismatch between electrons and phonons [14]), and if the Wiedemann-Franz law applies, equation 12 reduces to a simple asymptotic expression where $v_H \div J$:

$$v_H = [L / (1-T_b/T_c)]^{1/2} \ J / C_v \quad (13)$$

where L is the Lorentz number.

Hotspot speeds have been measured [16] on Nb, Nb_3Sn and NbN refractory materials. $v_H \geq 10^3$ m/s data for $J \approx J_c$ at 4.2 [K] agree well with eqn. 13.

<u>4) High-Resolution Position Detector</u>

A time-of-flight experiment [17] for the propagating hotspot [14,15,16], triggered by a particle event, is proposed for very high position resolution, corresponding to the slow speed v_H. In this experiment, two thermometers, for instance Giaever tunnel junctions, would be placed at each extremity of the wire as shown on Fig. 6.

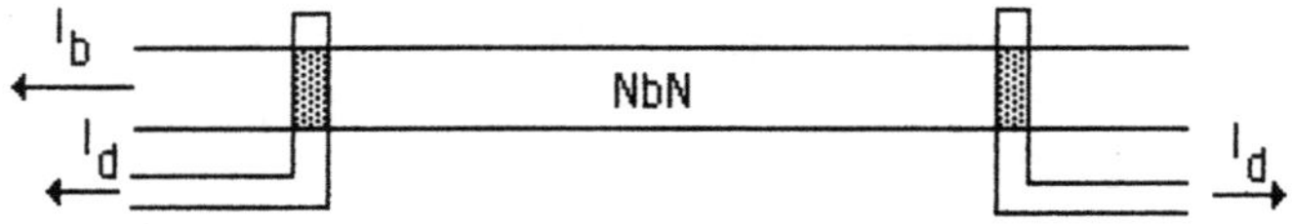

Fig. 6 : *Metastable* NbN *strip (driven at large enough current bias* I_b*) with thermometer junctions at each end (shaded zones) to detect the hotspot.*

Alpha-induced propagating hotspots have been observed [15] on In strips of width w = 34 µm, at relatively low temperatures $T_b < 3$ [K] and large current densities $J \geq 4.4 \cdot 10^5$ A/cm^2 ($W_J \geq 2 \cdot 10^5$ W/cm^3). NbN is suggested as a more suitable material than In : this compound can be made very resistive, leading to large Joule dissipation at lower current bias I_b. Moreover, close to the metal-insulator transition, NbN displays a slightly semiconducting resistivity behaviour above T_c. Combined with the high melting temperature, this property avoids fusing the wire ("burn-outs" of soft superconductors).

The two hotspot detecting junctions should be biased at low current I_d in the nonlinear part of their I-V characteristics, in order to detect the hotspot arrival at the junction as a voltage or resistance drop (see Figure 7). Notice that the whole detector can only be reset by dropping the current bias I_b of the wire ($W_J \ll W_M$ condition).

Since the hotspot speed is rather slow ($v_H \approx 10^3$ m/s), the proposed time-of flight experiment should lead to very high accuracy position determination in localizing the trigger particle event : a reading time accuracy $\delta t \approx 1$ ns only is

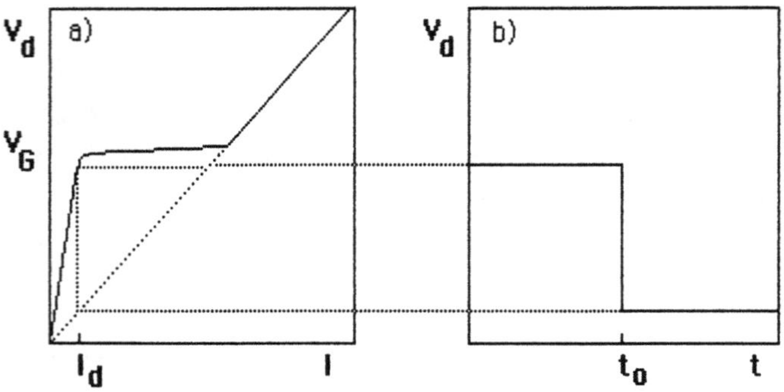

Fig. 7 : a) *I-V Characteristics of a hotspot detector Giaever junction. The junction acts as a switch between the superconducting state (continuous line) and in the normal state (dotted line) for the low reading current* I_d. b) *Detector voltage* V_d *at the hotspot arrival time* t_0. *The potential drop* V_d *between the* NbN *strip and the junction control line corresponds to the switch of the junction into the normal state.*

required for a position reading accuracy $\delta x \approx 1$ µm. This last feature is especially welcome for particles releasing low energies, leaving sub-micron primary hotspots. According to eqn. (9), a small hotspot diameter $D \geq 0.1$ µm is enough to trigger the propagating hotspot in a $w = 1$µm strip driven at the current $I_b = 0.9\, I_c$. Then, the strip width w can be matched to the position resolution $\delta x = w$. Even for NbN, the propagating hotspot condition $W_J >> W_M$ corresponds to these relatively high current densities.

The slow speed of this detector has also its drawbacks. Neglecting the reset time (leading to a factor ≤ 2 for $W_J = 0$ during the reset time), the strip count rate is $R \approx [v_H/L] = 10^5$ events[cm]/sec. However, the total count rate Γ of N independent channels is $\Gamma = \sum R = N \cdot R$ and becomes an intensive variable for large N if $w = \delta x$. Such a square detector consisting of N independent lines has $\Gamma = \delta t^{-1}$. In other words, Γ no longer depends on speed for such an array of detector lines. Details on such an N-line propagating hotspot will be published elsewhere.

6) Conclusion

The hotspot model is successfully applied to a wide range of superconducting particle detectors. Emphasis is put on experimental situations where "hot" quasiparticles ($T^* \geq T_c >> T_b$) are used as a trigger for metastable states. For switching Josephson junctions, this model predicts a detailed switching probability including the effect of fluctuations. The calculated current threshold and lifetime are in qualitative agreement with various experiments.

Application to particle-induced switching of strips is more involved, since the wire-resistive behavior depends on the current bias conditions. The model successfully predicts bolometer results with asymmetric pulses for low bias current densities. Moreover, for large bias current densities, the propagating hotspot condition applies, leading to a new detector application promising ultrahigh position resolution in the µm range and fast total count rates in the case of arrays of independent line detectors.

This work has been supported by the US-Department of Energy, BES-Material Sciences, under Contract # W-31-109-ENG-39.

References

1) See for instance K. E. Gray, in Nonequilibrium Superconductivity, Phonons, and Kapitza Boundaries, ed. by K. E. Gray (Plenum, New York, 1981), Vol. 65, p.p. 131-167.
2) G. H. Wood and B. L. White, Can. J. Phys. 51, 2032 (1973), and G. H. Wood and B. L. White, Appl. Phys. Lett. 15, 237 (1969)
3) D. Twerenbold, Europhys. Lett. 1, 209 (1986)

4) A. Barone, S. de Stefano, G. Darbo, G. Gallinaro, S. Siri, S. Vitale, and R. Vaglio, Proceeding of the LT-15 Conference, ed. U. Eckern, A. Schmid, W. Weber, H. Wühl (Elsevier Science Pub., B. V. 1984)
5) M. Tinkham, Introduction to Superconductivity (ed. McGraw Hill, New York, 1975)
6) A. Barone, S. de Stefano, and K. E. Gray, Nucl. Instr. and Meth. A235, 254 (1985)
7) R. Magno, M. Nisenoff, P. Shelby, A. B. Campbell and J. Kidd, IEEE Trans. Nucl. Sci. NS28, 3994 (1981); R. Magno, R. Shelby, M. Nisenoff, A. B. Campbell, and J. Kidd, IEEE Trans. Mag. MAG19, 1286 (1983)
8) A. Barone, and S. de Stefano, Nucl. Instr. and Meth. 202, 513 (1982)
9) H. A. Kramers, Physica 1, 284 (1940), and P. A. Lee, J. Appl. Phys. 42, 325 (1971)
10) T. A. Fulton and L. N. Dunkleberger, Phys. Rev. B9, 4760 (1974)
11) M. K. Wu, J. R. Ashburn, C. J. Torng, P. H. Hor, R. L. Meng, L. Gao, Z. H. Huang, Y. Q. Wang, and C. W. Chu, Phys. Rev. Lett 58, 908 (1987)
12) K. W. Shepard, W. Y. Lai, and J. E. Mercereau, J. Appl. Phys. 46, 4664 (1975)
13) D. E. Spiel, R. W. Broom, and E. C. Crittenden, Appl. Phys. Lett. 7, 292 (1965)
14) R. F. Broom and E. H. Rhoderick, Br. J. Appl. Phys. 11, 292 (1960)
15) D. E. Spiel, R. W. Broom, and E. C. Crittenden, Bull. Am. Phys. Soc. 9, 655 (1964)
16) K. E. Gray, R. T. Kampwirth, J. F. Zasadzinski, and S. P. Ducharme, J. Phys. F : Met. Phys. 13, 405 (1983).
17) M. Morgue and A. Gabriel, Rev. Sci. Instr. 55, 1000 (1984)

"Minicylinder" Design for Solar Neutrino Detection (A naive proposal)

G. Vesztergombi

Max-Planck-Institut für Physik und Astrophysik, Werner-Heisenberg-Institut für Physik, Föhringer Ring 6, D-8000 München 40, Fed. Rep. of Germany

1. Introduction

The main problem of solar neutrino or dark matter searches is to detect low-energy radiation with highest possible quantum efficiency. Due to the rather small cross-sections involved, it means monitoring a large bulk of material for small energy deposits.

The idea of using coherent neutral current scattering [1] can relax the multiton target mass requirement to the few kilogram level, but it implies to detect nuclear recoil energies in the 1 eV range. In principle, the superheated superconducting granule detector can fulfill these requirements; a number of practical problems should, however, be overcome before realization of this project [2].

Here I should like to present a new variant along these lines. Instead of granules, so-called "minicylinders" are proposed as basic elements providing the following advantages:

a) Target mass 1-2 kg is achieved by a reasonably manageable number of elements.

b) "Simple" mass production procedure(s) can provide standard, uniform detector elements

c) Optical read-out in parallel for thousands of channels

d) Background rejection by real 2-dimensional localization of the interaction point

Of course, there is a big unknown: what is the achievable sensitivity of this device?

2. Principle of Operation of a Superconducting Glass Rod Solenoid

Let us take a glass rod coated by some superconducting material (Fig. 1a) cooled down below $T < T_c$. Assuming $H\ (t = 0) \gg H_c$ magnetic field at time $t = 0$, no superconducting state is possible. Keeping the T temperature fixed by gradual decrease of the magnetic field at some t^* moment at field strength $H^* = H\ (t^*)$, the coating becomes superconducting, and additional drops of the outside magnetic field will be

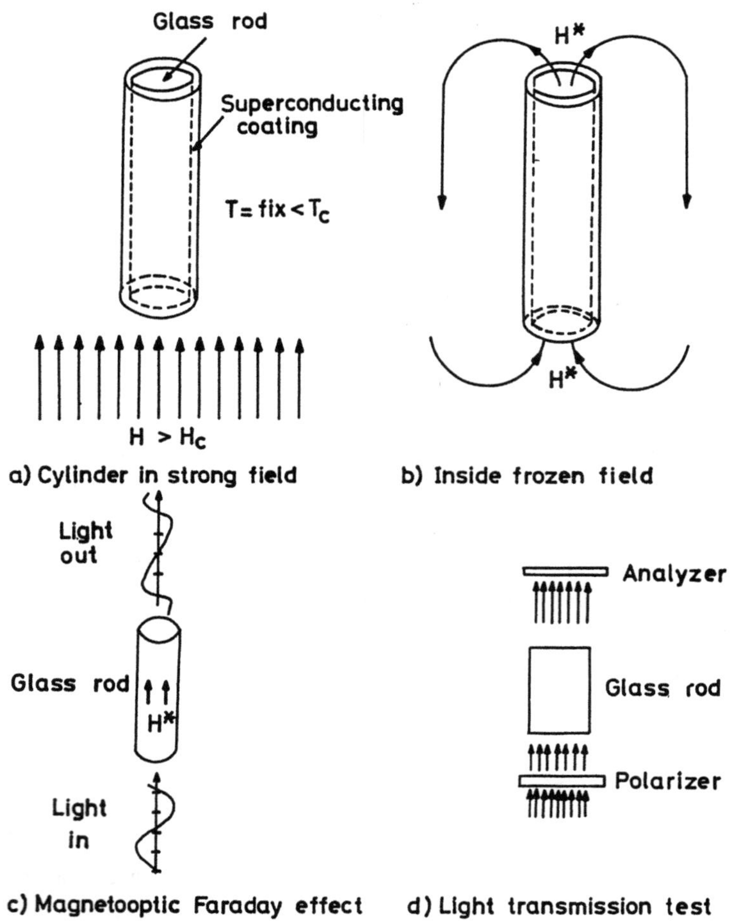

Figure 1. Principle of operation

compensated by the current induced on the cylindrical superconducting outer surface of the rod. That is, this H* field remains frozen inside the cylinder (Fig. 1b). In principle, one can play the same super-heating or -cooling game with cylinders as with the granules. The practical limits should be studied experimentally.

If one transmits plane-polarized light through the glass rod in presence of parallel magnetic field, the plane of polarization is rotated (Fig. 1c) by an angle Θ. This is the so-called magneto-optic Faraday effect. The amount of rotation is given by

$$\Theta = V \cdot H \cdot l \quad \text{where} \quad V = \text{Verdet-constant}, \; H = \text{field strength}, \; l = \text{rod length}$$

Some values for Verdet-constant are given in Table 1 taken from ref. [3]. One can put an analyzer into the way of transmitted light which is closed for light with

TABLE 1 Verdet Constant *V* for Faraday Rotation Glasses (after Snitzer, 1966)

Glass	λ, μm	*T*, K	*V*, minutes of arc/(Oe)(cm)	Useful λ, μm
Schott SFS-6 (Pb-Si)	0.700	300	0.071	1.45–2.0
	1.000	300	0.032	
Schott SF-6 (Pb-Si)	0.632	300	0.053	0.45–2.0
AO As-S	1.100	300	0.065	0.9–8
TeO_2, 20% PbO, 80%.	0.7	300	0.128	
Ce^{3+}-P	0.5	300	−0.326	0.4–2.0
		24	−2.57	
	0.7	300	−0.132	
Pr^{3+}-P	0.7	300	−0.123	0.65–0.85
Tb^{3+}-P	0.7	300	−0.150	0.5–1.4
Pr^{3+}-B	0.67	300	−0.243	0.65–0.85
Pr^{3+}-Al-Si	0.7	300	−0.198	0.65–0.85
		196	−0.655	
Tb^{3+}-Al-Si	0.7	300	−0.216	0.5–1.4
		196	−0.800	
Dy^{3+}-Al-Si	0.6	300	−0.272	0.5–0.67
		196	−1.03	

rotated polarization (Fig. 1d) but open for the incoming (original) polarization. If the coating is heated above the critical curve, it kills the superconducting current; therefore, the frozen H* field disappears, and the light will get through the heated cylinder.

The above theory is independent from the actual structure of the superconducting coating: it can be a continous surface or a dense coil made by winding 10 - 20 μ diameter insulated wire in one or more layers on the glass rod in such a way that the beginning and the end are joined together. In this case a small energy deposit at any point within the wire volume of ~ 10^3 - 20^3 μ^3 will stop the total current and kill the frozen magnetic field.

3. Glass rod Minicylinder Production

There is a straightforward way for production of glass rod solenoids. One should put the rod into a winding machine and solder the wire ends together. Only practice can decide whether the speed and quality is good enough. It is sure that this method should provide the first test samples for exploratory studies. It seems to be worth looking for alternative processes, too.

Here I should like to propose a method which resembles the chip-making processes. Instead of silicon crystal, one starts with a glass rod (Fig. 2a); by sputtering or other techniques, one produces a layer of 10 - 20 μ on the outer surface of the rod (Fig. 2b) and makes a photoresistant coating (Fig. 2c). Then by a well-focused

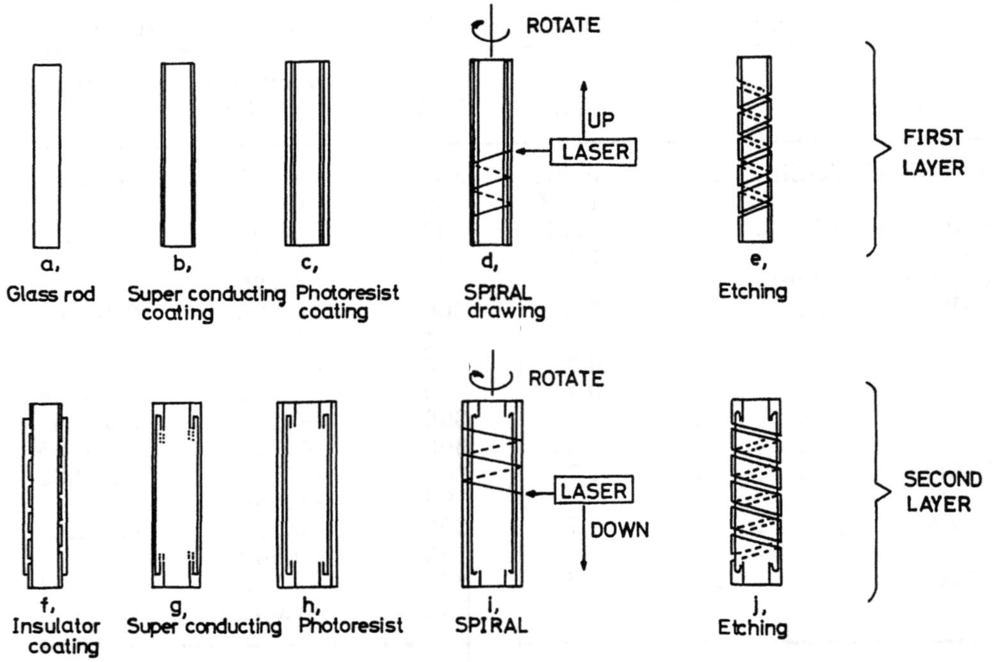

Figure 2. Minicylinder production

($< 1\ \mu^2$) laser beam, one can draw a SPIRAL on the rod, rotating it in such a way that the beam goes "up" by 10 - 20 μ during a full rotation (Fig. 2d); by etching away the SPIRAL, there remains one layer of ideally winded solenoid (Fig. 2e). Covering the surface by some insulating material, one can repeat the coil production process "downward" (Fig. 2f ...). The two coils will be joined together at their ends if one is careful enough not to put insulating material above the end circles of the first conducting layer.

This process seems to be very complicated but one can hope that by automatization finally it can gain in productivity.

4. Target Mass on a Single Rod

Due to the fact that only the superconducting material contained in the minicoils provides sensitive target, one should optimize the design by increasing the coil volume without losing sensitivity.

To achieve large enough optical rotation one needs several centimeters of glass. Let us take $l = 5$ cm. Diameter of $d \simeq 3$ mm would provide 1 cm circumference. Thus, the total surface of a single rod is 5 cm^2.

Due to coherent cross-section and material density consideration, one prefers high Z materials like Pb or W. Using 20 μ coating, one gets a volume for two layers on top of each other

$$V = 2 \cdot 5\ cm^2 \cdot 20\ \mu = 2 \cdot 10^{-2}\ cm^3$$

with mass $M_{Pb} = 0.228$ g and $M_W = 0.388$ g, respectively.

For additional optimization one can remember the scaling rule: decreasing the diameter by factor "k" one can pack into the same volume "k" times more target mass. Thus, the target mass density is inversely proportional to the rod diameter. Pushing for "microcoils" becomes important only if the cryostat volume turns out to be more expensive than the rod production price. There should be an optimum because the number of rods in a given volume is increasing as k^2.

Another way of increasing target density is to put more than 2 layers connected serially on the same rod.

5. Detector Assembly

The essential elements of the setup are shown in Fig. 3. The rods can be arranged into a hexagonal matrix. In our example of 3 mm rod diameter, a hexagon with 30 cm diameter contains 7651 rods providing

$$M^{TOT}_{Pb} = 7651 \cdot 0.228 = 1.74\ kg,\ M^{TOT}_{W} = 2.97\ kg$$

target mass, respectively.

The so-called MASK-IN and MASK-OUT masks before and after the rod-matrix assure that the light can be transmitted only through the glass rods. The CCD-matrix of

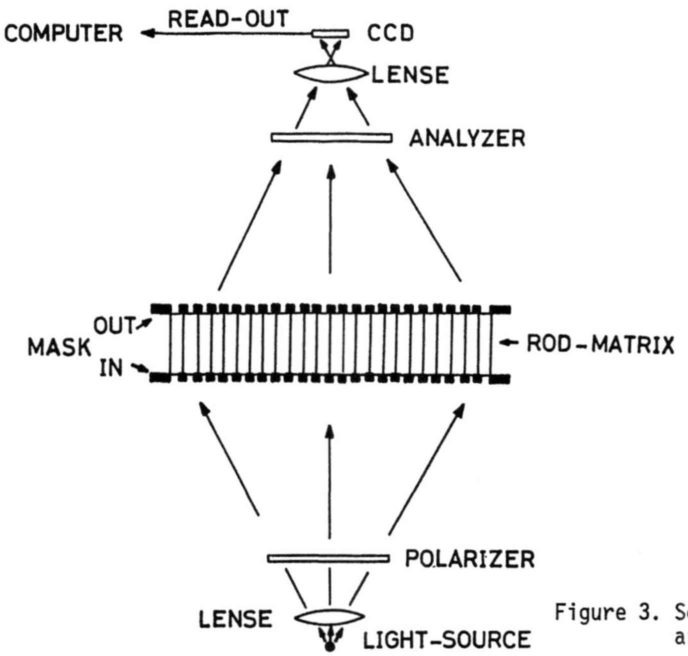

Figure 3. Schematic layout. Crystostat and magnet not shown

256 x 256 size can cover the image of the total surface with great redundancy because a single rod will extend to about 6 - 8 pixels.

The light source is flashed in regular time intervals, determined by the readout speed of CCD. Computer can compare the last and previous picture. Trivial algorithm can detect whether the magnetic field has changed inside any rod.

The setup can work in superheated regime like the granules when the strong field from outside of the rod enters into it, or in a simple resistance flip regime when the outside field is zero and the trapped field is escaping in consequence of the neutrino interaction with the superconducting layer.

References

1. A. Drukier and L. Stodolsky: Phys. Rev. D 30, 2295 (1984)
2. K. Pretzl these proceedings
3. E. Hartfield and B.J. Thompson: In Handbook of Optics, ed. by Driscoll, Vaughan, (Mc Graw-Hill, New York, 1978) 17-21

Electron Beam Detection with Superheated Superconducting Grains

A. de Bellefon, P. Espigat, D. Broszkiewicz, and R. Bruere-Dawson

Laboratoire de Physique Corpusculaire, College De France,
F-75231 Paris Cedex 05, France

Superconducting Indium grains have been proposed as a solar neutrino detector /1/.

As a matter of fact among possible targets to detect low energy neutrinos, Indium is the only nucleus to have some unique properties and allowing a real time electronics experiment /2/. The capture reaction is :

$$\nu_e + {}^{115}In \longrightarrow {}^{115}Sn + e^- + \gamma_1 + \gamma_2$$

Such an experiment is under feasibility study /3/. In this communication we want to present the work we have done to determine physical constraints on a detector and our experimental results concerning the detection of an e^- beam.

1 - CONTRAINTS

One of the most important features of an Indium experiment is the intrinsic background due to spontaneous In beta decay. Accidental coincidence of time -uncorrelated background can fake a solar neutrino capture event. For instance two beta decays, or a beta which fakes the electron and a stray γ -ray around 600 KeV which fakes the whole delayed event.

The rejection we need has been estimated : if we assume 5.10^{14} yr as a life time for Indium we then have 2 decays /s/ cm³ (In) to compare with a rate of 10^{-11} evt /s/ cm³ from solar neutrinos.

Such a big rejection factor is achieved with the help of a rather good spatial resolution together with some energy resolution /4/. The detector has to be divided into a number of cells. With a reasonable segmentation the spatial resolution could stay around 1 cm³. Then, for a trigger on capture electron and the two photons we have to get an energy resolution of 50 KeV around 100 KeV and 100 KeV around 500 KeV to reach a level of signal over noise ratio of the order of 10.

Are these numbers realistic ? To answer that question a Monte Carlo simulation computer program has been run.

Monte Carlo Simulation

Our program is divided in three sections :

1/ Generation of the medium microspheres embedded in paraffin with a given filling factor - Generation of events.

2/ Tracking of the particle through that medium : Energy losses and multiple scattering for an electron - Interaction of the photon with matter.

3/ Taking into account the diamagnetism effect evaluation of flipping probability for one grain under the ionising particle energy deposit.

100 electrons of a given energy have been generated. We have got a gaussian distribution for the number of grains struck by the electrons. The number is expressed as a function of E $n = f \pm (E) = \langle n \rangle \pm \sigma_n$ from these results an energy resolution is obtained. $\sigma_E = \frac{dE}{d\langle n \rangle} \sigma_n$

Results for σ_E are listed below for E = 100 KeV

Operating T	1 K	0.2 K
Size		
5 μ	σ_E = 100 KeV	σ_E = 50 KeV
2 μ	σ_E = 25 KeV	σ_E = 15 KeV

At low temperature and for small grain, these numbers are very close to those we need to reject the background.

So in principle since we can read and obtain small grains it is possible to envisage the following steps to study feasibility.

2 - EXPERIMENTAL DETAILS

In our experiment Tin superconducting grains are used as the target of an electron beam produced by a high voltage Van de Graaf accelerator.

A magnetic field, parallel to the beam, is applied by means of a conventional solenoid able to deliver a field up to 500 gauss. Sample itself is cooled down in a cryostat as shown on Fig. 1.

2 - 1 Samples

Specimens of tin microspheres have been prepared in Paris (G.P.S.) by sonic dispersal of the metal in a suitable oil. Tin grains thus obtained were dried and selected according to size by flitration through calibrated sieves. Grains were then embedded into paraffin wax and installed in a twenty turns loop as shown on Fig. 1 below.

For each sample we obtained the superheating cycle CURVE.

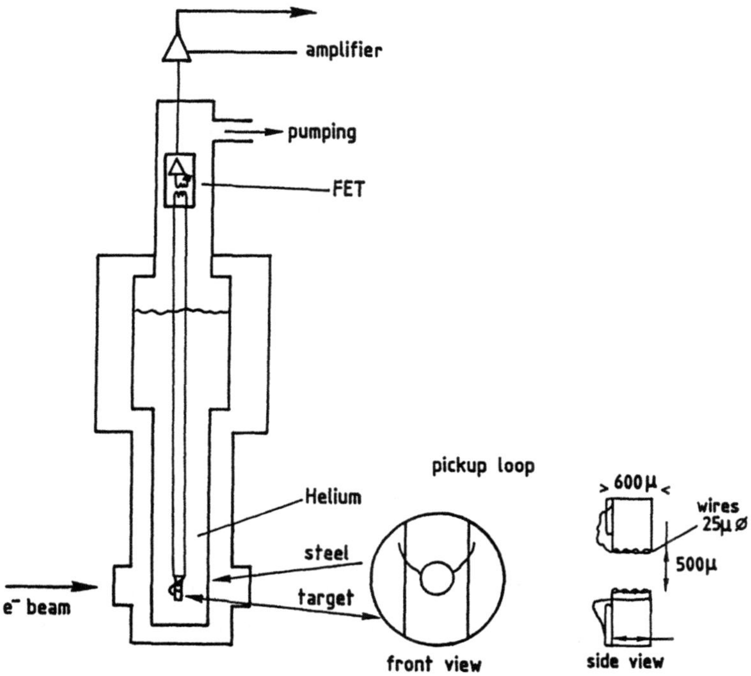

Fig. 1

2 - 2 Cryogenics

The experimental set up is shown on Fig. 1. A ^{4}He reservoir keeps our superconducting sample at 4.2 K and then by pumping we can go down 1.5 K. The helium reservoir is surrounded by vacuum tank. The electron beam goes through an Aluminum window at the end of the beam pipe, a mylar window at the entrance of the cryostat and 100 μm of stainless steel at reservoir wall.

2 - 3 Electronics - Read out - Amplifier

Grains embedded in a paraffin wax are installed in a pick up coil as shown on Fig. 1. Up to now performances of such a detector have been studied by measuring the resonance frequency of a LC Circuit. Since two years we are able to couple the flux variation or the voltage pulse to the room temperature electronics by means of a transmission line whose characteristics are presented on Fig. 2.

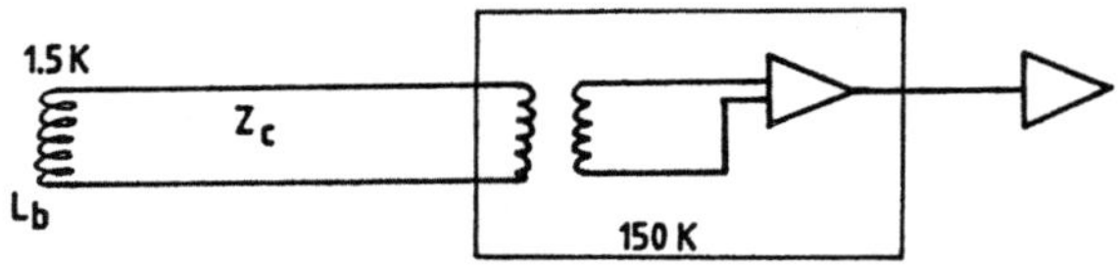

Fig. 2

Details on the amplifier and the shape of the signal have been given in another talk /5/.

Briefly said we work with the following typical parameters :

L_b/Z_c = Tr (amplifier rise time)

Tr = 170 ns

Zc = 300 Ω

L_b = 150 nH

3 - IRRADIATION

All previous studies or experiments /6/ have been done with an LC Circuit. A direct pulse detection of magnetic field occurs in a short time Δ t, during which a grain produces a flux variation $\Delta\Phi$ in the pick up loop surrounding it. We read $\Delta\Phi/\Delta t$ a voltage pulse in the μ V range. The amplifier we used has a good enough sensitivity to read grains below 10 μ diameter in a loop of 0.5 mm diameter. We used a sample of tin grains below 10 μ diameter with a 20 % filling factor.

First we characterized the sample by sweeping the field without beam from o to H_{sh}, and indeed seeing grains flipping individuallly, derivative of the superheating curve Fig. 3a.

Next the field is raised from o to H_o, and stabilized at H_o. The electron beam at minimal intensity is switched on during 10 seconds, and switched off. The cycle is then completed by sweeping from H_o to H_{sh}.

The difference between the two cycles gives the number of grains which has been flipped by the beam at Ho. On fig. 3b we can see that all nearly metastable grains have flipped. From Fig. 3c, we can estimate Δ H which would have been necessary to flip these grains.

This corresponds to a temperature shift $\Delta T = \frac{\Delta H}{H_{SH}} \times \frac{T_c^2}{2T}$

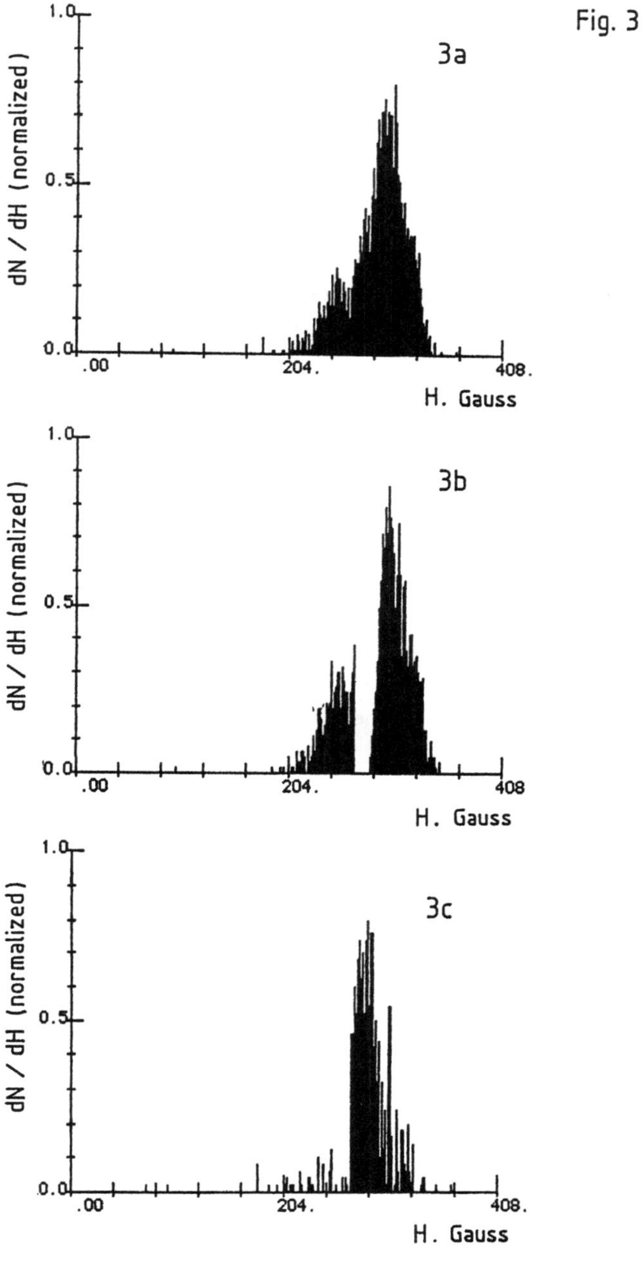

T is the operating temperature,

T is the critical temperature

Hsh is the superheating field at O°

The corresponding amount of energy deposited is given by $\Delta E = C_p V \Delta T$ where V is a typical grain volume.

In our sample for the observed Δ H we have calculated $\Delta E \geq 50$ KeV which is more than the maximum energy left by a 2 MeV electron in one grain.

Our explanation is that 10^{10} e^- (1 n A) produce 10^7 bremsstrahlung photons with energy around 40 or 50 KeV. These photons are produced in the steel wall. They peak forward in the beam direction. At low energy and because of the energy distribution they are reabsorbed inside the wall and at high energy they are very few to start with. So grains are flipped mainly by bremsstrahlung photons.

As a matter of fact low energy photons interact by photoelectric effect, and the photoelectrons lose their energy inside one grain and are much more efficient than a minimum ionising electron.

4 - CONCLUSION

For the first time an electron beam effect on superconducting grains has been observed by direct pulse read-out amplification method. But, mainly we saturate our detector. That means : bremstrahlung photons overflow electron effect. Now we have to see electron individual effect and measure its energy.

To do that, we need first to reduce the beam flux down to thousands of e^- and to trigger on the electrons which have gone through the grains. This will be our next stage.

Discussions with Mr. Froissart have been very much appreciated as well as his critical point of view.

1/ G. Waysand : Proceedings of the Moriond Meeting on Massive neutrinos in Astrophysics and in Particle Physics (1984).

2/ R.S. Raghavan, Phys. Rev. Letters 37, 259, (1976).

3/ Faisabilité de la détection de neutrinos solaires avec l'Indium. Internal Report, April 1987.

4/ A. de Bellefon, P. Espigat, G. Waysand AIP, Conf. Proc. 126 1984) N. Booth same Proceedings.

5/ G. Waysand in Proceedings of this workshop.

6/ Bernas et Al. Physics Letters 24 A, 721, (1967) J. Blot, T. Pellan, J.C. Pineau, J. Rosenblatt Journal of Applied Phys. 45, 1429, (1974).

7/ D. Hueber, C. Valette, G. Waysand, Cryogenics, 387, (1981).

Monte Carlo Simulation of a Double-Beta Decay Experiment with Superconducting-Superheated Tin Granules

P. Andreo[1], J.G. Esteve[2], and **A.F. Pacheco**[2]

[1]Sección de Fisica, Hospital Clinico Universitario, E-50009 Zaragoza, Spain
[2]Departamento de Fisica Teórica, Facultad de Ciencias, E-50009 Zaragoza, Spain

The double beta process [1] is a rare but very important nuclear decay mode which exists in some tens of even-A nuclei, where the relative position of the energy states of the isobars is abnormal, i.e. the usual β decay $(A,Z) \rightarrow (A, Z \pm 1)$ is energetically forbidden -or very hindered because of selection rules-, leaving the $\beta\beta$ transition $(A,Z) \rightarrow (A, Z \pm 2)$ as the only allowed way of decaying. In the nuclear tables one may find up to 26 cases of $\beta^-\beta^-$ parents [2]. Their decay, if the neutrino is a Dirac particle ($\nu \neq \bar{\nu}$), leads to the daughter nuclei plus two electrons and two antineutrinos in the so called two-neutrino double beta decay ($\beta^-\beta^-_{2\nu}$). If the neutrino is a Majorana fermion ($\nu = \bar{\nu}$), and hence the lepton number L is not conserved, then in addition to that mode the neutrinoless one ($\beta\beta_{0\nu}$) is also possible, and in this case the two ejected electrons carry all the energy released in the process.

The appeal of this phenomenon which can be traced to the phenomenological implications of Grand Unified Theories and the question of the neutrino mass has triggered a considerable experimental development [1]. Primigenic ores from different parts of the world have been analyzed by geochemical methods [3] providing for the first time, experimental evidence for the $\beta^-\beta^-$ process in ^{82}Se, and ^{130}Te. Under a similar philosophy a radiochemical method [4] has also been suggested for ^{238}U. The direct methods have the advantage of discriminating between the two $\beta\beta$ modes by measuring the energy sum of the two electrons. This can be accomplished in a cloud chamber [5]; or, recently, in a time-projection chamber [6]. Another type of direct measurement specially suited to detect the $\beta\beta_{0\nu}$ mode [7] is based on ^{76} Ge, which has the great advantage that germanium is an excellent detector of electrons, so that the source is also the detector [8].

In spite of the remarkable progress achieved, the number of $\beta\beta$ isotopes analyzed so for is modest, and its enlargement would be very useful. In fact, the search for other experimental techniques is valuable in itself and it could help to shed light on the remaining puzzles, such as the theory-experiment controversy about the half-lives of the two-neutrino mode. With that aim in mind I suggested the possibility of using superconductivity for ββ detection [9]. Here, we continue the exploration of that idea, doing a Monte-Carlo simulation of the performance of a detector in which microgranules of a ββ substance (which is at the same time a class 1 superconductor), are embedded in paraffin [10]. The temperature and external magnetic field are chosen to fix the sample in a superheated state so that small deposits of energy produced by the outgoing ββ electrons trigger a thermal nucleation and are able to flip some granules to the normal conducting state. The quick disappearance of the Meissner effect induces a voltage pulse in a surrounding coil which is the handle of detection. We see, hence, its close similarity with the case of the neutrino-indium proposal, so popular in this workshop [11]. Among all the $\beta^-\beta^-$ isotopes, only 10 of them are superconducting substances. A first selection, using the criterion of Q >2 MeV, gives the following nuclei: ^{96}Zr, ^{100}Mo, ^{116}Cd and ^{124}Sn. Our MC simulation has been done with Sn, but that is not to be considered as a final option, and indeed our conclusions are useful whatsoever be the chosen nucleus. Obviously, to do a firm choice, the background presented by each isotope has to play a basic role. Among the $\beta^+\beta^+$ isotopes with superconducting properties, ^{96}Ru and ^{106}Cd are the clear favorites, because they are the only ones able to release two positrons.

Two features to be positively noted in a ββ detector based on superheated granules are: its small volume because the source and the detector are the same thing; and the spatial resolution obtained by the grid of coils, which may be a very useful tool to reject non ββ events. The main drawback lies in its presumed poor energy resolution, which handicaps ***ab initio*** its use as a $\beta\beta_{0\nu}$ detector, because the high energy tail of the $\beta\beta_{2\nu}$ mode produces exactly the same type of signal. In any of the two ββ modes, one can think of the advantage of looking at the $0^+ \rightarrow 2^+$ transition [12], ending in the first excited state of the daughter nucleus, by detecting the two electrons and the subsequent characteristic gamma ray. In our simulation we will, however, analyze only the $0^+ \rightarrow 0^+$ transit.

The mode with two neutrinos is certainly less "sexy" than the other but very important from the nuclear structure point of view. Here one is probing a double Gamow-Teller matrix element

$$M_{GT}^{2\nu} = \langle 0_F | \sum_{n,m} \tau_n^+ \tau_m^+ \, (\vec{\sigma}_n \cdot \vec{\sigma}_m) \; | 0_I \rangle \qquad (1)$$

which has proven to be very difficult to be safely computed. In fact there is still a striking discrepancy in the value of the half lives for ^{82}Se. The theoretical prediction [13,14] is a factor of 10 shorter than the experimental result [3,6]. For ^{130}Te and ^{128}Te the discrepancy reaches up to a factor of 100 [3, 13, 14] To calculate the order of magnitude of a $\beta\beta_{2\nu}$ detector based on ^{124}Sn, we will suppose that $T_{1/2}$ (^{124}Sn) = $6*10^{19}$ years. This is a mere [1] phase-space estimation [15] taking the nuclear physics factor $|M_{GT}2\nu / \mu|^2 = 10^{-2}$. Taking into account the 5.98% of isotopic abundance of ^{124}Sn, one needs 17cm^3 of natural tin to have 1 event per day. So that if the dilution factor of tin in paraffin is 20%, the total volume needed is 85 cm^3, i.e. a cube with sides of 4.4 cm. This scale of few centimeters contrasts sharply with the size necessary in realistic neutrino experiments.

To have the neutrinoless mode, as is well known [1], the Majorana nature of the neutrino has to be completed by a non-zero neutrino mass (m_ν) and (or) an explicit chirality impurity in the leptonic current (η) which induces a departure from a pure V-A theory. In the hypothesis of $\eta = 0$, the decay width is proportional to m_ν^2 and to the modulus square of a nuclear matrix element, which for light neutrinos is

$$M_{GT}^{0\nu} = \langle 0_F \mid \sum_{n,m} \tau_n^+ \tau_m^+ \; (\vec{\sigma}_n \cdot \vec{\sigma}_m) \; |\vec{r}_n - \vec{r}_m|^{-1} |0_I\rangle \quad (2)$$

The great performance of detectors based on germanium has provided increasingly longer experimental bounds for $T_{1/2}^{0\nu}$ (^{76} Ge), up to a present value[17] of 3.4 * 10^{23} years; which has pushed m_ν to values of the order of few eV. The exploration of a competitive halflife in ^{124}Sn of say $T_{1/2} = 10^{23}$ years, assuming the same dilution factor as before, would demand a total volume of 1.4 * 10^5 cm^3.

In our simulation the detector is assumed to be a cube of 10 cm^3 volume, consisting of Sn spheres of a diameter ϕ, immersed in a paraffin environment. With a 20 % dilution factor, the Sn granules are symmetrically distributed in the detector volume and centered in elementary cubes. In order to account for differences in the size of manufactured granules, a random disperson of a 50% around the central value ϕ_0 of the diameter of the grains has been considered

The EGS4 Monte Carlo system [18] as modified at NRCC [19,20] has been used to simulate the transport of electrons, positrons and bremsstrahlung photons, generated either in the $\beta\beta_{2\nu}$ or $\beta\beta_{0\nu}$ decay of ^{124}Sn. The energy deposited in each Sn granule is scored during the simulation and compared with the critical energy of the phase transition to determine the total number of spheres that modify their metastable state in each event.

Some details of this transport code are as follows. Knock-on electrons can be created and followed down to a cut-off energy of 1 KeV. 1Kev electrons have a continuous slowing down approximation (csda) range in Sn orders of magnitude smaller than the value of ϕ used in this work, so that this is a reasonable asumption for the energy cutoff. Due to the small dimensions considered here, there are also some reasons based on the scattering of charged particles to choose a limit as low as possible that will be commented later on. The same cut-off is used for the transport of bremsstrahlung generated photons. The simulation of Rayleigh scattered photons is included in the simulation.

In order to properly account for the small dimensions of the grains, the maximum allowable charged particle step-size was set equal to the minimum distance between two nearest spheres. The influence of the maximum allowable fractional energy loss of charged particles step[19] in Sn was investigated by letting this parameter vary between 4 and 0.1%. For paraffin, a fractional energy loss of 1% was chosen due to the similarity of this material to water regarding energy absorption and scattering, where this value produces very accurate results for low energy electrons. As a check of energy conservation through the simulation, the total energy deposited in Sn and paraffin, as also the energy of escaping photons, is scored separately and compared with the initial kinetic energy of both electrons (K). In the neutrinoless mode K is alway equal to To = 2.288 MeV, i.e the Q value of the process. In the two-neutrino mode the probability density fuction for K is

$$d\Gamma^{2\nu}/dK \propto K(T_0-K)^5\,[K^4+10\,K^3m+40K^2\,m^2+60Km^3+30\,m^4] \qquad (3)$$

where m is the rest mass of the electron. It presents a maximum for about one third of T_0. For a given value of K, the kinetic energy of the first electron T_1 is sampled from a probability density function

$$d\Gamma\,/\,dT_1 \propto (T_1+m)^2\,(T_0-T_1+m)^2 \qquad (4)$$

This is obtained from the $\beta\beta_{2\nu}$ kinematics and is coincident with the $\beta\beta_{0\nu}$ case (if η=0). The velocity of the first electron is obtained from T_1 and isotropic directional cosines. The second electron is initiated at the same position as the first one, with an energy T_2=K-T_1, and oriented according to the angular correlation function

$$C\,(\vec{p}_1\cdot\vec{p}_2) = 1-(\vec{p}_1\cdot\vec{p}_2)\,(T_1+m)^{-1}\,(T_2+m)^{-1} \qquad (5)$$

for both modes (when η=0).

Due to the high number of spheres in the detector it is unrealistic to account for all the granules at the same time to produce a three-dimensional distribution of the energy deposited in the whole detector. Preliminary runs were done in order to determine the maximum number of granules where energy is deposited in each event. This allowed the posterior inclusion of a "counter" for each "touched" granule with a considerable reduction of the computation time for event.

An important part of the code is made of routines handling the geometry of the problem. Due to the small dimensions of the spheres, electron tracks cross many Sn-paraffin boundaries, and pathlengths have to be continuously verified in order to keep each segment of the electron path within a homogeneous region where energy losses and scattering characteristics are the same. For this purpose, some of the standard routines included in the EGS4 distribution package were modified accordingly. The closest allowable distance to a surface was of the order of the csda range of an electron with energy equal to the cut-off. Double precision was used for all the energy and distance computer variables .

It has to be mentioned that due to the dimensions handled in this work and the restrictions on step-size and energy loss for step, multiple scattering of charged particles is not accurately considered in the simulation. This is a limitation in the Moliere's theory itself, where there must be a minimum number of colisions (≥ 20) for the theory to be valid [21]. The restriction imposes a lower limit to the distance travelled by an electron with a given energy, that for tin must be superior to $3\beta^2$ µm approximately. Charged particles with kinetic energies greater than about 0.5 MeV will then have their multiple scattering switched off by EGS4 for the small step-sizes chosen here. On the other hand, and due to the low cut-off energy used in the simulation, a large number of electron - electron inelastic collisions will be individually considered, and the kinematics of the process will exactly predict the changes in the direction of both electrons. The balance between small step-sizes and low energy cut-off was then considered to be the best possible compromise considering the high symmetry of the detector.

During the simulation the three coordinates of any touched granule, the energy deposited there, and the specific value of the diameter of that sphere are scored. Due to the small size of the spheres, we have confidently assumed a global heating model for the change of state of the granules. Then, the threshold energy E_{th}, necessary to produce the flip, obtained from equilibrium thermodynamics, is the sum of two volume terms, i.e. one from the specific heat of phonons and other from Cooper pairs, and a surface term [22]. Specifically we have.

$$E_{th} = \Delta Q_{vol}(1) + \Delta Q_{vol}(2) + \Delta Q_{surf}$$

$$\Delta Q_{vol}(1) = N\alpha\,(\Delta H/H)^2\,T_c^4$$

$$\Delta Q_{vol}(2) = N\,2b\,a^{-1}\,\gamma T_c^2\,(\Delta H/H)\ \exp[-a/\ 2\Delta H/H] \qquad (6)$$

$$\Delta Q_{surf} = .05\ \ 2\ H_c(0)^2\,\xi_0\ R^2\,(\Delta H/H)$$

N is the number of moles in the granule, and the rest of parameters for Sn are: a=1.41; b=7.63; T_c= 3.70^{0}K ; H_c(o)= 306 gauss; α= .242 mJ/mol $^0K^4$; γ=1.80 mJ/mol $^0K^2$; ξ_o= .23 μm. (ΔH/H) is a measure of the relative distance to the superheated line, which can vary from one granule of the sample to another. We have assumed it varies in the range [22].

$$.005 \le (\Delta H/H) \le (2\sigma/300\ \text{gauss}) \qquad (7)$$

according to the central part of a gaussian function of mean deviation equal to σ. σ in realistic experiments has a value around 30 (gauss). Using (6) with ΔH/H sampled from a restricted gaussian, those granules where the energy deposit is bigger than E_{th} are selected at the end of the simulation. Our results for the number of flipped granules (n) as a function of the average diameter of the sample appear in Table 1.

TABLE 1. NUMBER OF FLIPPED GRANULES

ϕo (μm)	σ = 30 gauss		σ = 20 gauss
	0v	2v	2v
1.	296.4	100.9	113.3
2.	58.5	21.6	25.2
3.	21.7	7.9	9.9
4.	10.8	3.8	4.9

There, one observes the fast decay of n as ϕ_0 is increased, which proves the necessity of using very small granules in this type of experiment. If σ is decreased to 20 gauss, there is a corresponding increase in n because E_{th} is decreased in average. The information on the spatial spreading of the flipped granules is needed to be able to distinguish events from background. In Fig 1 we have plotted how n is distributed around the point of decay for ϕ=1μm and σ=30 gauss. As indicated there, squares and triangles stand for the $\beta\beta_{0v}$ $\beta\beta_{2v}$ modes respectively.

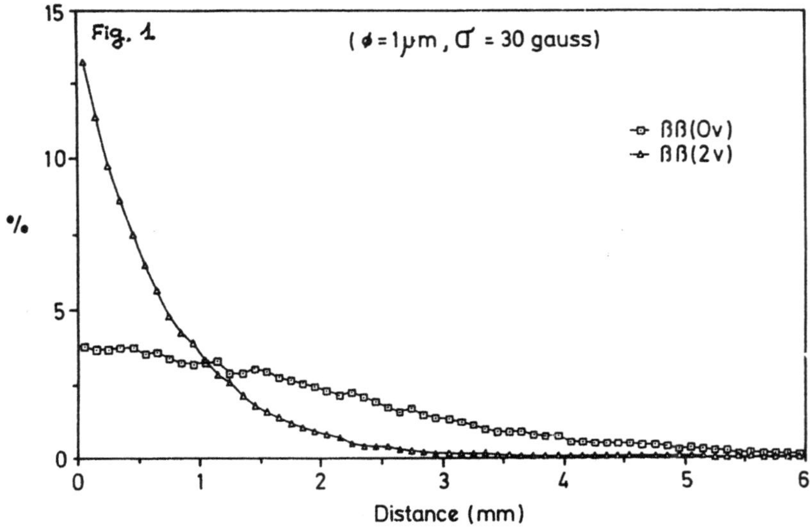

Each point represents the number of flips in a spherical shell of 100μm of width. The density distribution is expressed as the percentage (%) of the total number n. We see that in the two-neutrino mode all the granules are flipped practically within a sphere, centered around the decay point, of radius not bigger than 1mm. The neutrinoless mode on the contrary produces transition of granules up to say 3 or 4 mm. This type of percentage plot has a remarkable stability as a function of ϕ_0, as we see in Fig.2 and Fig.3, where the two modes are presented independently. The results obtained for the range of flipping show that a detector in the scale of a few centimeters, with a spatial resolution of say 0.5 mm^3, would be enough to distinguish these events from other interactions.

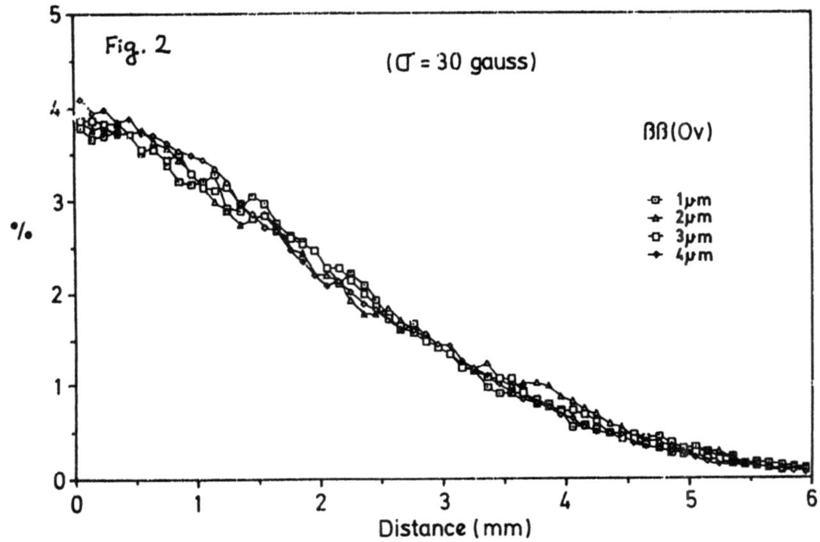

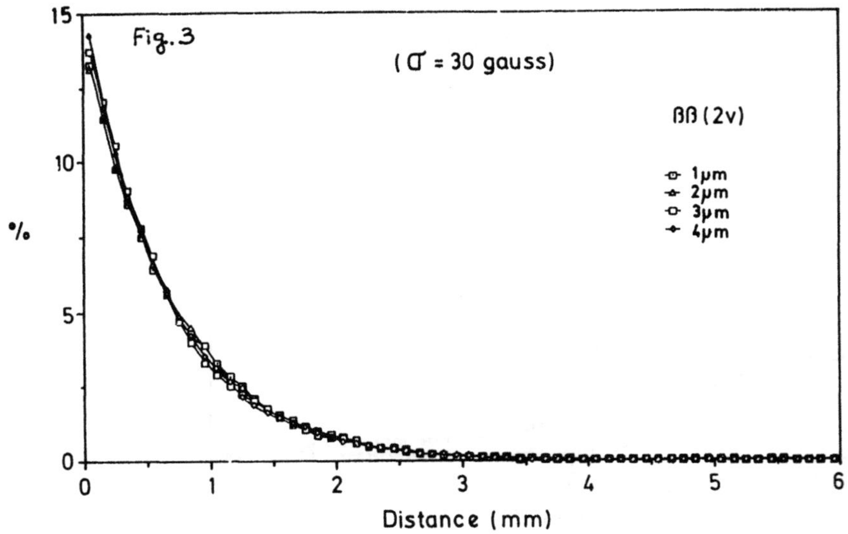

This work is a first step, which should be completed with a simulation of the electromagentic signal produced, using a specific array of coils, plus a detailed study of the most likely sources of background. In this sense, preliminary analysis shows that cadmium looks like a more clean substance than tin [23]. These studies will be continued in the future.

A. F. P. would like to acknowledge very useful discussions with L.González-Mestres and D. Perret-Gallix, during a visit to CERN. That visit was supported by a grant of the Programa Europa de Estancias de Investigación CAI-DGA. Comments by D.W.O. Rogers and A.F. Bielajew are also acknowledged by P.A.

REFERENCES

[1] For recent reviews see for example T. Kotani: "Double Beta Decay and Majorana Neutrino". Prog. Theor. Phys. Suppl. , No. 83 (1985), or W.C Haxton,G.J. Stephenson Jr.: "Double Beta Decay". Progress in Particle and Nucl. Phys. Volume 12, 409 (1984).

[2] E.Fiorini: La Rivista del Nuovo Cimento 2, 1 (1972).

[3] T. Kirsten, H. Richter and E.K. Jessberger: Z. Phys. C16, 189 (1983).

[4] W.C. Haxton, G.A. Cowan and M. Goldhaber: Phys. Rev. C28 467 (1983).

[5] M. Moe and D. Lowenthal, Phys. Rev. C22 2186 (1986).

[6] S.R. Elliott, A.A. Hahn, and M.K. Moe: UCI Preprint (FEB 86/, unpublished).

[7] E. Fiorini et al, Phy. Lett. 25B, 602 (1967).

[8] E.Bellotti, et al, Phys. Lett. 146B, 450 (1984).
[9] A. F. Pacheco: Mod. Phys. Lett. 1, 167 (1986).
[10] H. Bernas, J.P. Burger, G. Deutscher, C. Valette, S. Williamson, Phys. Lett. 24A 721 (1987).
[11] G. Waysand: these proceedings.
[12] A. Molina and P. Pascual: Nuovo Cimento A41, 756 (1977).
[13] K. Grotz and H.V. Klapdor: Phys. Lett. 142B, 323 (1984).
[14] W.C. Haxton, G.J. Stephenson and D. Strottman: Phys. Rev. D25, 2360 (1982).
[15] In a microscopic calculation GROTZ and KLAPDOR [16] obtained $T_{1/2}^{2\nu}$ (^{124}Sn) = $9.3 * 10^{19}$ years.
[16] K. Grotz and H.V. Klapdor: Phys. Lett. 157B, 242 (1985).
[17] D.O. Caldwell, et al, Phys. Rev. Lett. 54, 281 (1985).
[18] W. R. Nelson, H. Hirayama, D.W.O. Rogers : SLAC Report 265 (1985).
[19] D.W.O. Rogers : Nucl. Instr. Meth. 227, 535 (1984).
[20] D.W.O. Rogers, A.F. Bielajew : "The use of EGS for Monte Carlo Calculations in Medical Physics". Report PXNR-2692 (1984). National Research Council of Canada.
[21] G. Moliére : "Theorie der Streeung schneller geladener Teilchen II. Mehrfach und Vielfachstreuung" . Z. Naturforschog 3a, 78 (1948).
[22] L. González-Mestres, D. Perret-Gallix: "New Results on the Basic Properties of Superheated Granules Detectors". LAPP-EXP-86-05, January (1987)
[23] L.González-Mestres: private communication.

An Indium Solar Neutrino Experiment

N.E. Booth[1], D.J. Goldie[1], B.M. Hawes[1], D.A. Hukin[1*], J.L. Lloyd[1], C. Patel[1], G.L. Salmon[1], J.E. Evetts[2], J.H. James[2], J.M. Lumley[2], G.W. Morris[2], and R.E. Somekh[2]

[1]Department of Nuclear Physics, University of Oxford, Keble Road, Oxford OX1 3RH, U.K.

[2]Department of Metallurgy and Materials Science, University of Cambridge, Pembroke Street, Cambridge CB2 3QZ, U.K.

We report on our progress during the past three years in developing a superconducting indium neutrino detector. Two highlights are the quasiparticle trapping mechanism, and the merger of technologies to provide a viable route for fabricating stable detector elements.

1. Introduction

Solar neutrinos can in principle be detected via three different processes:

(a) neutrino-electron scattering

$$\nu_e + e^- \rightarrow \nu_e + e^-$$

where the recoil electron is detected,

(b) neutrino-nucleus coherent scattering

$$\nu + (Z, A) \rightarrow \nu + (Z, A)$$

where the nuclear recoil is detected (there is also the recently suggested process of nuclear excitation by neutral-current scattering [1]),

(c) inverse β-decay

$$\nu_e + (Z, A) \rightarrow (Z + 1, A) + e^-$$

where the nucleus $(Z + 1, A)$ is identified radiochemically, or the process is detected in real-time.

Proposed detectors based on the first two reactions suffer from the lack of a clear signature for neutrino interactions. They may be considered as potential detectors of ^{8}B neutrinos, but not at the present time as detectors of neutrinos from $p - p$ fusion. In contrast, the inverse β-decay reaction on ^{115}In [2] has a very low threshold (128 keV) and a very high sensitivity to $p - p$ and ^{7}Be neutrinos. It also has a very characteristic signature.

2. The ^{115}In Reaction

Apart from the low threshold, ^{115}In has some unique properties. It is the only nucleus with which a real-time electronic experiment can be performed to measure the differential neutrino energy spectrum. The reaction is

$$\nu_e + {}^{115}\mathrm{In} \rightarrow {}^{115}\mathrm{Sn}^* + e^-,$$

* Present address: Crystalox, Wantage, Oxon. OX12 9AJ, U.K.

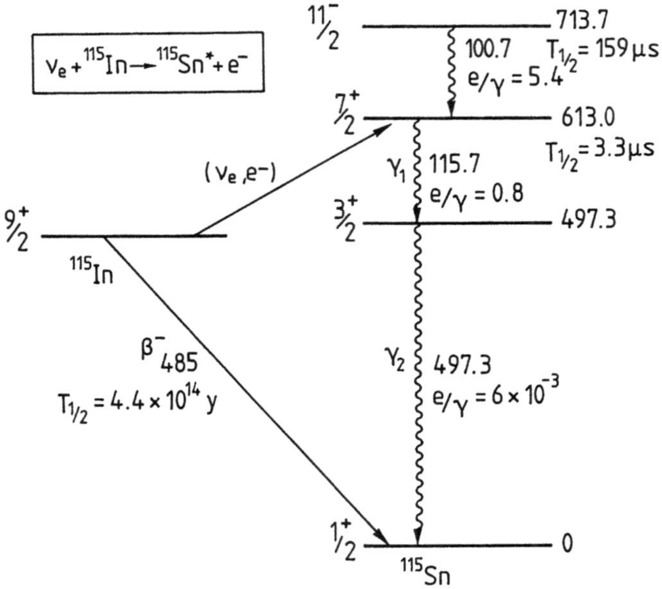

Fig.1. The ^{115}In neutrino capture reaction

and the relevant energy levels of ^{115}In and ^{115}Sn are shown in Fig.1. Neutrino capture with electron emission takes ^{115}In (96% natural abundance) to the second excited state of ^{115}Sn. The unique signature for neutrino detection is a pulse from the electron, followed by, on average 3 μs later, two time-coincident pulses, one spatially very close to the electron with energy 116 keV, and the second of 497 keV. The energy of the electron is $E_e = E_\nu - 128$ keV, so a measurement of its energy gives the neutrino energy. The expected electron energy spectrum based on the Standard Solar Model [3] is shown in Fig.2 plotted in bins 20 keV wide. This model predicts a

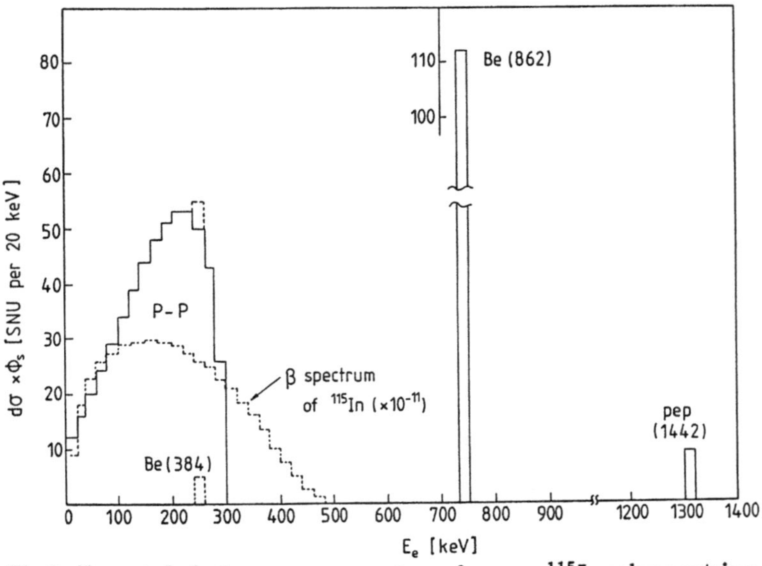

Fig.2. Expected electron energy spectrum from an ^{115}In solar neutrino experiment, plotted in bins 20-keV wide: also shown is the β-decay spectrum of ^{115}In converted to SNU and multiplied by 10^{-11}

rate of 1 event/day for 3.3 tonnes of natural indium. Therefore for a 1 tonne detector running for 1 year we expect 90 events in the $p-p$ distribution, 18 events in the ^{7}Be peak, and 1 or 2 in the $p-e-p$ peak. For a detector with good energy resolution it is worthwhile to look in detail at the ^{7}Be peak. Its width depends upon the central temperature of the Sun and is expected to be about 1 keV.

Neutrinos will also interact with an indium target via neutral current processes. Although these processes give higher rates, the only signature is a small energy deposition which is extremely difficult to select from background in any detector.

Although the ^{115}In reaction has a very characteristic signature, the basic problem in a low-rate experiment is to select the signal and reject the various backgrounds. Another problem is to get the indium in such a form that the detector is sensitive to the low energy (average 200 keV) electrons from the neutrino capture and the 116 keV γ-ray. The detector must be designed to make as full use as possible of the unique signature of the ^{115}In reaction.

There are several possible backgrounds, the most obvious being accidental time- and energy-coincidences due to the natural radioactivity of ^{115}In. In Fig.2 we have also plotted the β energy spectrum multiplied by 10^{-11}. Eighty per cent of the β's overlap the energy range of electrons from neutrinos from $p-p$ fusion. Therefore, we must make full use of the very characteristic delayed time-coincidence signature of the ^{115}In reaction. The signature can be recognized above background by dividing the detector into 10^4 or more segments, by having moderately good time resolution between γ_1 and γ_2 and good energy resolution. We also note that γ_1 travels a mean distance of only 0.76 mm in pure indium before it interacts, and 93% of the time deposits its full energy in the first interaction. On the other hand γ_2 travels a mean distance of 13.5 mm and only 20% of the time deposits its full energy in the first interaction. In the other 80% of the interactions it gives up only part of its energy to an atomic electron by the process of Compton scattering. Subsequently it can undergo further Compton scattering but eventually deposits its remaining energy via the photoelectric effect in the same or a neighbouring segment.

We have recently made careful background estimates based on a model of a realistic detector [4]. The results show that the accidental coincidences can be handled provided the ambient γ-ray background is not higher than that achieved in recent experiments on double β-decay, and provided the detector has good energy resolution, comparable with that achievable with Ge detectors. Therefore we are developing a new type of detector—one which uses indium in its superconducting state.

3. Superconducting Indium Detectors

Indium is a superconductor with a transition temperature $T_c = 3.408$ K. When metals are in the superconducting state the valence or free electrons near the Fermi level (energy E_F) are bound in Cooper pairs (binding energy $2\Delta(T)$) due to the special nature of the interactions between electrons and phonons. At any temperature $T < T_c$ Cooper pairs are occasionally broken by phonons, and the excitations, called quasi-electrons and quasi-holes, or collectively quasiparticles, can recombine and emit a phonon. The number-density of quasiparticles $N(T)$ is proportional to the Boltzmann factor $\exp(-\Delta(T)/kT)$ and so is strongly temperature dependent. The recombination rate follows a similar dependence.

What happens when an ionizing particle, such as an electron from a neutrino interaction, deposits its kinetic energy in a superconductor? In the first 10^{-12} s many electrons are excited to energies well above E_F. Within this time they lose energy by interacting with other electrons and by radiating numbers of phonons. Those phonons with energy $\Omega \geq 2\Delta(T)$ interact mainly by breaking Cooper pairs [5]. After about 10^{-10} s there remains an excess of quasiparticles, of order 10^9 per MeV of energy deposition, due to Cooper-pair breaking, and an excess of phonons of energy $\Omega < 2\Delta(T)$. Compared with a similar ionization event in a semiconductor, a superconductor has about 10^3 times more electronic excitations, and a larger fraction of the energy in the electronic system instead of in phonons or lattice excitations [6].

The best way to detect such an excitation is to use a superconducting tunnel junction. This consists of the detecting film or crystal, a tunnel barrier consisting of a thin insulating layer (often the natural oxide), and a counter-electrode film of the same or a different superconductor. We [7] and other groups [8–12] have made such devices using superconducting thin films, and detected pulses due to α-particles and X-rays. Because of the smallness of the energy gap parameter $\Delta(T)$ the energy resolution should be 30–50 times better than that of semiconductor particle detectors. In fact our first results extrapolate to an energy resolution of better than 0.1 keV at 500 keV. This is a factor of 10 better than we assumed in the background calculations [4].

To do the solar neutrino experiment we must have the indium in pieces of bulk material, at least 1 g each, and not in the form of thin films. Moreover we need sufficiently good time resolution to resolve the electron emitted in the neutrino capture process from the delayed 116-keV γ-ray which will almost always interact in the same piece. Further, we need to solve the technical problems of making junctions on the surface of bulk indium, electrical contacts, etc.

In Fig.3(a) let S_1 represent a piece of indium of volume V with a junction of area A_J on one surface. We will assume that the volume of the counterelectrode S_3 is negligible compared to that of the single crystal detector S_1, so that all the ionizing events of interest break Cooper pairs in S_1. For simplicity we assume however that S_3 is the same superconductor as S_1. An ionizing event at $t = 0$ creates n_0 excess quasiparticles via Cooper-pair breaking. These excess quasiparticles although created locally will diffuse rapidly in high-purity single crystal material,

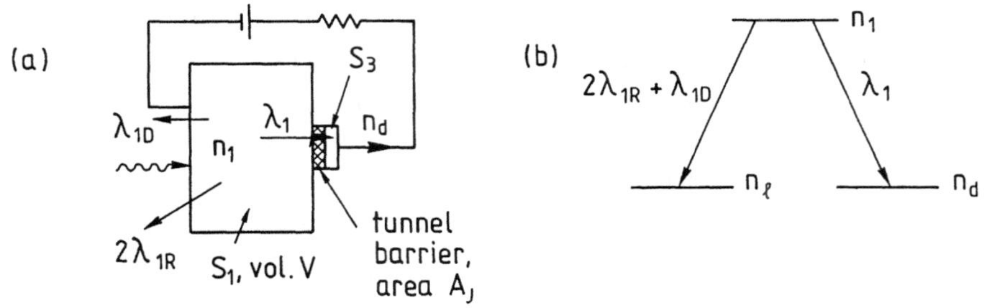

Fig.3.(a) Schematic arrangement of a tunnel junction on the surface of a superconducting crystal of volume V (b) Radioactive decay analogue

and after a few transit times across the crystal will be uniformly distributed throughout the volume V. For the present discussion we will not consider the signal current during these first few transit times since it depends on the location of the ionizing event, on the geometry, and on the quasiparticle diffusion coefficient, and requires a detailed calculation. With these simplifications the operation of the detector may be envisaged by analogy with the radioactive decay scheme shown in Fig.3(b). Here $n_1(t)$ is the excess number of quasiparticles in S_1, $n_d(t)$, the number which have tunnelled (rate constant λ_1) and $n_l(t)$ the number which have been lost due to recombination (rate constant $2\lambda_{1R}$, the factor 2 being due to the fact that 2 quasiparticles are lost in a recombination event) and due to out-diffusion into the electrical lead to S_1 (rate constant λ_{1D}). The differential equations corresponding to Fig.3(b),

$$\frac{dn_1}{dt} = -(\lambda_1 + \lambda_{1D} + 2\lambda_{1R})n_1 \tag{1}$$

and

$$\frac{dn_d}{dt} = \lambda_1 n_1 \tag{2}$$

are easily solved to give the signal current

$$i_s = |e|\frac{dn_d}{dt} = \lambda_1 n_0 \exp(-\Lambda_1 t) \qquad (3),$$

where

$$\Lambda_1 = \lambda_1 + 2\lambda_{1R} + \lambda_{1D} \qquad (4).$$

The total charge that can be collected is

$$q(\infty) = |e| n_0 \lambda_1/\Lambda_1 \qquad (5).$$

For good operation we want $\lambda_1 \gg \lambda_{1D}, \lambda_{1R}$. The out-diffusion rate can be minimized by geometrical considerations of the electrical contact to S_1. We will show later how it can be totally eliminated. The recombination rate can be made arbitrarily low by operating at a low temperature, but care must be taken to avoid any normal-metal regions or magnetic impurities.

The tunnelling rate constant λ_1 is given by [13]

$$\lambda_1 = \frac{1}{2e^2 N_1(0)}\frac{G_{nn}}{V} \qquad (6),$$

where $N_1(0)$ is the single-spin density of states at the Fermi surface of S_1 ($13.6\times10^{21}\,\mathrm{eV^{-1}\,cm^{-3}}$ for In) and G_{nn} is the normal state conductance of the junction. Unfortunately the signal current is inversely proportional to the volume of the crystal. Making A_J large increases the junction capacitance C_J ($\simeq 4\mu\mathrm{F\ cm^{-2}}$ for indium oxide) and this increases the signal rise-time. The factor to optimize is the specific conductance g_{nn}. The highest value achieved experimentally is about $10^8\,\Omega^{-1}\,\mathrm{cm}^{-2}$. Inserting this and $A_J = 1\,\mathrm{mm}^2$ into (3) gives, for $n_0 = 10^9$ (about 1 MeV energy deposition) in a crystal of volume 1 cm^3, $i_s = 0.04\,\mu$A. Assuming we have a load $R_L = 50\,\Omega$ and that the dynamic resistance of the junction, R_D, is much larger than R_L we expect a maximum signal voltage of 2 μV. This is very small, but is detectable with a reasonable signal-to-noise ratio with low-noise semiconductor electronics. However, the latter sets a limit on the energy resolution which can be achieved. Moreover, the high value of g_{nn} which we have used in the calculation may not be routinely achievable.

As a result of thinking about how to circumvent this problem we came up with the idea of quasiparticle trapping [13]. Suppose that between the indium crystal and the junction we have a small volume v of another superconductor S_2 (in practice aluminium) which has a smaller gap than indium. Over the distance of a few superconducting coherence lengths the indium gap Δ_1 will change to the aluminium gap Δ_2 as shown in Fig.4. Then quasiparticles created in the indium can diffuse into the aluminium where they can lose energy by phonon emission and drop down to the gap edge of aluminium. They now have insufficient energy to get back into the indium and are trapped in the smaller volume of aluminium where they hit the tunnel barrier much more frequently, thereby increasing the signal current. The additional rate constants λ_2 (for trapping), and λ_3 (for tunnelling out of the trap) are introduced in Fig.5(a). Figure 5(b) shows the radioactive decay analogue. The trap opens up another decay channel which can be much faster than the direct process with rate constant λ_1. We have also allowed for recombination in S_2 via λ_{2R}. How well such a device works depends upon the scattering time $\tau_s(E/\Delta_2)$ for a quasiparticle of energy E ($\approx \Delta_1$) in S_2 to relax to close to the gap edge of S_2 by phonon emission. The energy dependence is known theoretically [14] and several measurements have been made for aluminium [15,16]. Some of the results are shown in Fig.6. From these results we can estimate $\tau_s = 5$ ns for the indium-aluminium combination.

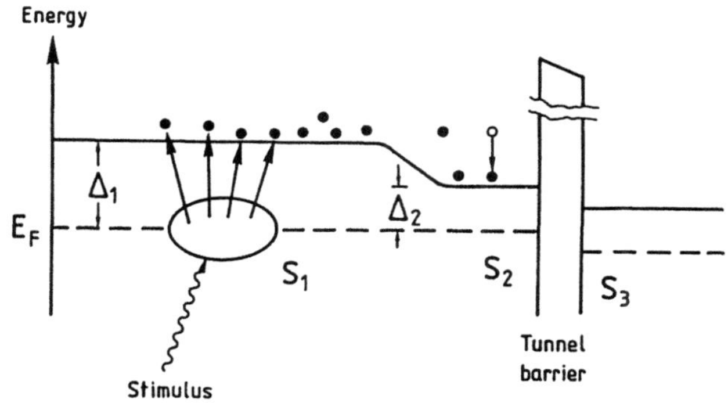

Fig.4. Energy diagram showing trapping of excess quasiparticles produced in superconductor S_1 by an adjacent superconductor S_2 of lower gap

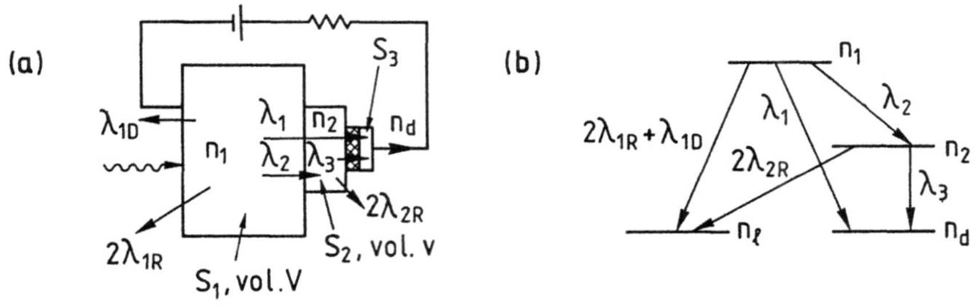

Fig.5.(a) Schematic arrangement of a detector S_1 incorporating a trap S_2 (b) Radioactive decay analogue

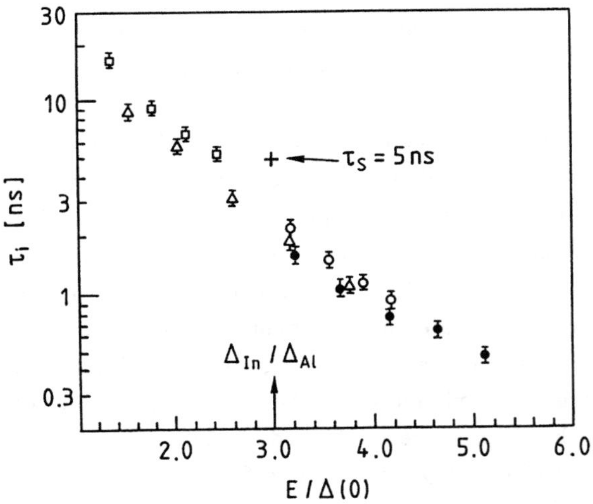

Fig.6. Some experimental results by MOODY and PATERSON [16] on quasiparticle relaxation in Al. The results, after correction for recombination, permit an estimate of $\tau_s = 5$ ns for the scattering time for an In detector with an Al trap

Assuming that the fraction of time the excess quasiparticles spend in S_2 is just the volume fraction $v/(V+v) \simeq v/V$ we have

$$\lambda_2 = \frac{v}{V}\frac{1}{\tau_s(\Delta_1/\Delta_2)} \tag{7}$$

and

$$\lambda_3 = \frac{1}{2e^2N_2(0)}\frac{G_{nn}}{v} \tag{8}.$$

Neglecting λ_{2R}, the optimum situation occurs for $\lambda_3 = \lambda_2 \gg \lambda_1$, which gives

$$\left(\frac{v}{V}\right)^2 = \frac{\tau_s(\Delta_1/\Delta_2)}{2e^2N_2(0)}\left(\frac{G_{nn}}{V}\right) \tag{9}.$$

Figure 7(a) shows the current pulse shape for $G_{nn}/V = 10^6\Omega^{-1}$ cm^{-3} and various trap volumes. The signal current can be increased by a factor of more than 100. Figure 7(b) shows how the maximum signal current and the optimum trap size depend upon G_{nn}/V.

When the ratio of the gaps Δ_1/Δ_2 is greater than 3, as for example Ta or Nb with an Al trap, a multiplication process can occur because some of the phonons emitted in the relaxation process can break Cooper pairs in S_2, thus increasing the number of quasiparticles in the trap. This leads to the concept of the quasiparticle multiplier, a device which can act as a low-noise superconducting preamplifier to further increase the signal current [13].

As mentioned previously most tunnel junctions have been fabricated by evaporating a metal strip onto a substrate, oxidizing the metal, and then evaporating another strip to form the counterelectrode. Unless very special alloys are used, for example the Pb-In-Au alloy developed by IBM for the base electrode, such junctions do not stand up to thermal cycling. In fact some cannot be stored in room-temperature air for more than a few hours.

Moreover, the fabrication of junctions on bulk single-crystal surfaces is much more difficult. The major problem is that the surface, even if recently chemically polished, is not as good as a fresh vacuum-evaporated film on a carefully cleaned substrate. The other problems arise from the necessity of insulating the counter-electrode contacts from the crystal, the possibility of

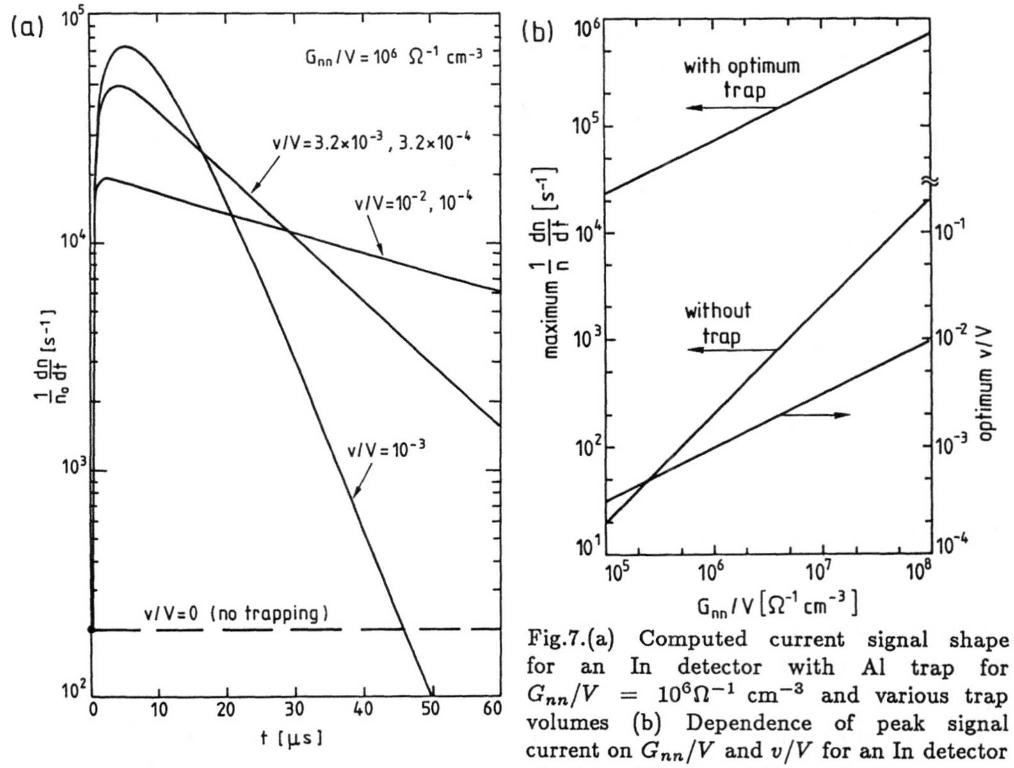

Fig.7.(a) Computed current signal shape for an In detector with Al trap for $G_{nn}/V = 10^6\Omega^{-1}\ cm^{-3}$ and various trap volumes (b) Dependence of peak signal current on G_{nn}/V and v/V for an In detector of volume V and an Al trap of volume v

damaging the insulation when attaching wires, and, in the case of soft metals like indium, the need to mechanically mount the crystal and connecting wires without damaging the crystal.

For the growing of crystals we have found that the vertical Bridgman technique with a graphite mold always yields good single crystals. However, we found that, whatever the mold material, if it was highly polished so as to give a smooth surface to the crystal, the crystal almost invariably stuck to it. We were forced to use a rough surface of graphite, and chemically polish the crystals afterwards. We were then faced with solving the many problems mentioned previously.

Meanwhile work in Cambridge was focused on the technology of the whole-wafer processing method of fabricating refractory metal tunnel junctions [17]. In this process a high quality film of Nb is sputtered on to one surface of a sapphire substrate which has been specially cut to give a good crystal-lattice match to the Nb. In the same vacuum cycle a very thin (~ 2 nm) layer of Al is sputtered on, and this is then oxidized by introducing a low pressure of oxygen into the deposition chamber. It is well known that a very stable, tenacious oxide forms on Al under these conditions. Then another thicker layer of Al is deposited, and finally another layer of Nb, all films over the entire surface. This results in a tunnel junction over the whole surface, as shown in Fig.8(a). After this several photoresist patterns are used and unwanted material is etched away by ion-beam etching and reactive-plasma etching. The latter process is highly selective;

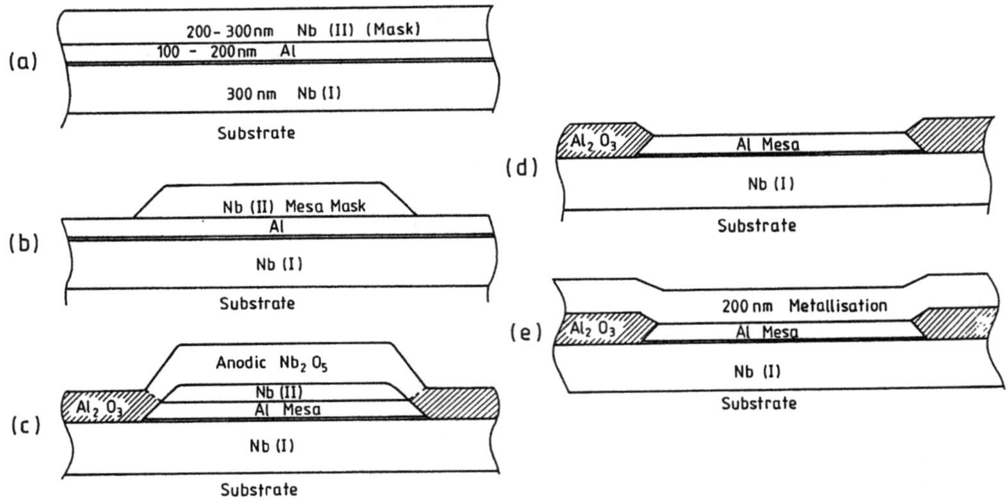

Fig.8. Schematic outline of the initial processing steps for fabricating indium detectors

Nb and its oxides are readily etched by the products of the decomposition of CF_4 gas in an r.f. discharge, whereas Al and its oxide are completely resistant. The result is indicated in Fig.8(b). In another step the thicker layer of Al is liquid-anodised to form an insulating layer of aluminium oxide which also defines the areas of the tunnel junctions. The final step, shown in Fig.8(e), is to deposit a metal film which is patterned to produce the contacts to the counterelectrodes. In contrast to almost all other methods of fabrication, in this method the tunnelling barrier remains buried and undisturbed in all the subsequent processes after its initial formation. This method, together with the metallurgical considerations which result in the choice of the right materials, yields robust, high-quality tunnel junctions.

But what, if anything, has this to do with junctions on single crystal indium? The answer is—everything!—it is the missing link. Armed with the knowledge that indium sticks to almost any smooth surface, we just grow an indium crystal over an existing tunnel junction structure which has been fabricated on a sapphire substrate. A photograph of one of our first successful attempts is shown in Fig.9. In practice it is only slightly more complicated than this. We have found that the process of crystal growing has no adverse effects on the junctions.

Note that the second layer of Al was used to produce an anodized insulating layer everywhere except over the desired area of the junction. We now have a layer of Al between the junction and the indium crystal. This is by design and is our first step in implementing the trapping idea. Our next step will be to increase the volume of the trap to its optimum value.

4. Plans for a Full-Scale Experiment

One may well ask the questions—is it feasible to extrapolate this technology to a full scale experiment; what about growing 10^5 crystals; can one cool one tonne of indium, etc.? These problems may seem formidable, and at the very least not trivial.

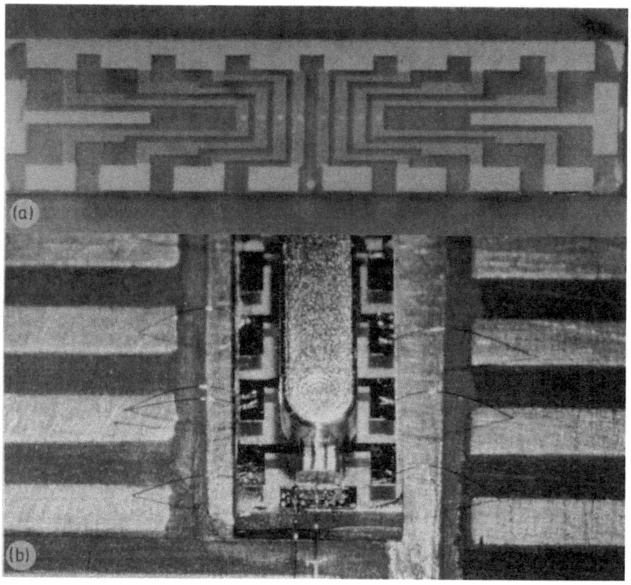

Fig.9. The merger of two technologies resulting in indium detectors (a) device fabricated on a sapphire wafer prior to final metallisation. The white dots are Al mesas which are joined to the peripheral Nb contact pads by Nb strips covered with insulating anodic oxide, (b) detector element, shown end on, after growth of an indium crystal making contact with the Al mesas, and after wire-bonding to Cu contact strips in a test rig.

Let us start with the detector technology. Most of the fabrication steps are very similar to steps used in the mass-production of semiconductor devices. Sapphire wafers are being extensively used in the new silicon-on-sapphire technology. The procedures shown in Fig.8 can be implemented on whole wafers. We can envisage several indium crystals with mass of 100 g or more on a wafer. We have already grown strings of crystals of this sort of mass in a number of geometries. It is important that the mass of sapphire be much less than the mass of indium because of the considerations discussed in Section 2. Mechanical supports for the wafers will also be of low mass.

Arrays of wafers will be in a bath of liquid ^{4}He in thermal contact with the mixing chamber of a large ^{3}He–^{4}He dilution refrigerator, or the bath may be the mixture itself. Cooling such a large mass of material is not a major problem. The main problem is minimizing heat input due to the wiring to the detectors and due to radiation.

A schematic diagram of what a full-scale experiment might look like is shown in Fig.10. The actual detector consisting of up to a few tonnes of indium will be less then 1 m^3 and relatively easy to shield against backgrounds. Detector elements of materials other than indium could be incorporated as "dark matter" detectors.

Perhaps the most difficult problem is to find enough expert manpower and money to bring the development to fruition and make a working experiment.

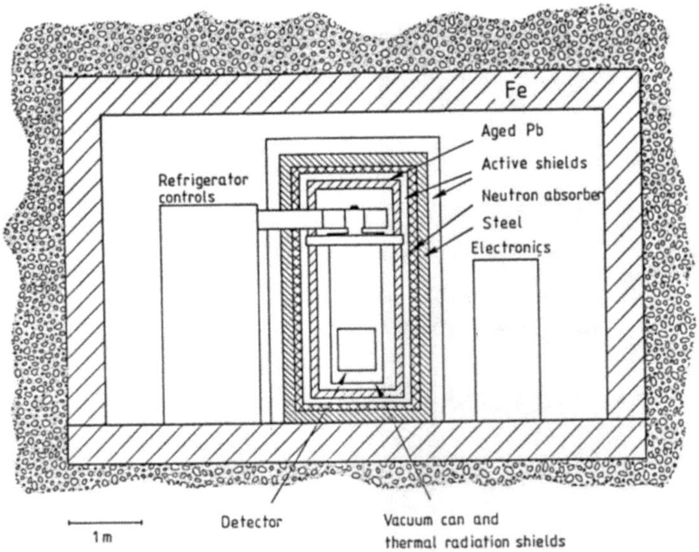

Fig.10. Schematic drawing of a several-tonne indium solar neutrino experiment

Acknowledgments

We have benefited from discussions with D.F. Moore on device fabrication and J.C. Barton on backgrounds. Fruitful discussions with A. de Bellefon, J.P. Maneval, J. Rich, M. Spiro and G. Waysand are gratefully acknowledged.

References

1. R.S. Raghavan, S. Pakvasa and B.A. Brown: *Phys. Rev. Lett.* **57**, 1801 (1986)
2. R.S. Raghavan: *Phys. Rev. Lett.* **37**, 259 (1976)
3. J.N. Bahcall: *Rev. Mod. Phys.* **50**, 881 (1978); J.N. Bahcall, W.F. Huebner, S.H. Lubow, P.D. Parker and R.K. Ulrich: *Rev. Mod. Phys.* **54**, 767 (1982); J.N. Bahcall, B.T. Cleveland, R. Davis, Jr. and J.K. Rowley: *Astrophys. J.* **292**, L79 (1985)
4. N.E. Booth: Oxford Nuclear Physics Laboratory Report, Ref: **19/87** (1987); N.E. Booth: *Sci. Prog. Oxf.* (submitted for publication)
5. J.J. Chang and D.J. Scalapino: *J. Low Temp. Phys.* **31**, 1 (1978)
6. M. Kurakado and H. Mazaki: *Nucl. Instrum. Methods* **185**, 141 (1981)
7. N.E. Booth, G.L. Salmon and D.A. Hukin: in *Solar Neutrinos and Neutrino Astronomy*, ed. by M.L.Cherry, W.A.Fowler and K.Lande, *AIP Conf. Proc.* No.**126**, (American Institute of Physics, New York, 1985) p.216
8. G.H. Wood and B.L. White: *Can. J. Phys.* **51**, 2032 (1973)
9. M. Kurakado: *J. Appl. Phys.* **55**, 3185 (1984)
10. A. Barone, G. Darbo, S. de Stefano, G. Gallinaro, A. Siri, R. Vaglio and S. Vitale: *Nucl. Instrum. Methods* **A234**, 61 (1985)
11. H. Kraus, Th. Peterreins, F. Pröbst, F. v. Feilitzsch, R.L. Mössbauer, V. Zacek and E. Umlauf: *Europhys. Lett.* **1**, 161 (1986)

12. D. Twerenbold: *Europhys. Lett.* **1**, 209 (1986)
13. N.E. Booth: *Appl. Phys. Lett.* **50**, 293 (1987)
14. S.B. Kaplan, C.C. Chi, D.N. Langenberg, J.J. Chang, S. Jafarey and D.J. Scalapino: *Phys. Rev.* **B14**, 4854 (1976); **B15**, 3567(E) (1977)
15. C.C. Chi and J. Clarke: *Phys. Rev.* **B19**, 4495 (1978)
16. M.V. Moody and J.L. Paterson: *Phys. Rev.* **B23**, 133 (1981)
17. J.M. Lumley, R.E. Somekh, J.E. Evetts and J.H. James: *IEEE Trans. Magn.* **MAG–21**, 539 (1985); J.H. James: paper presented at Europhysics Study Conference on Tunnelling at Low Temperatures, Leuven, Belgium, 1985

Cryogenic Detection of Particles, Development Effort in the United States

B. Sadoulet

University of California, Berkeley, CA 94720, USA

We review the development of cryogenic detectors of particles in the USA, with emphasis on large mass devices. Most groups are still tooling up and exploring basic properties of sensors. We summarize the main discussion themes and describe some of the early experimental results.

1. PHYSICS MOTIVATIONS

1.1 Potential Properties of Cryogenic Detectors

It has been realized recently [1] that it may be advantageous to detect particles through processes different from ionization. The break-up of Cooper pairs in a superconductor (leading to formation of quasi particles) or the generation of phonons in a crystal provide several interesting properties.

-Detectors sensitive to these processes would be sensitive in the bulk and would be made quite massive (several hundred grams). Some methods may allow to implement good position sensitivity.

-Since the detected energy quanta is much smaller (1meV for quasi particles, 10^{-6}eV for phonons at 10mk) that electron-ion or electron-hole pairs (20 eV and 1 eV respectively). It may be possible to obtain very low thresholds and very good energy resolution.

-As explained under section 2.1.5 the combination of phonon and ionization detection may allow to give a signature that an interaction occurred on a nucleus,[2] providing in some cases a much needed rejection of the radioactive background [3].

Because these detectors have to work at very low temperature, they are usually referred to as **cryogenic detectors.**

1.2 Physics Applications

The above properties have attracted a large number of groups. (See table I for a tentative list of groups developing large mass cryogenic detectors in the USA).

The most developed application is the use of bolometer (i.e. phonon detectors) for **X ray astrophysics** [1c]. The goal is to obtain a resolution of 1 eV for each photon.

Cryogenic detectors are also quite attractive for searches of **dark matter** if it is made out of particles (see e.g. Ref. [3] and references therein). They may complement in an essential way [4] the current searches [5] using conventional technologies. This is the main motivation in the USA for the effort of the UC Berkeley / LBL team and of Drukier et al.

Table I
Development of large mass cryogenic detectors in the USA

Group	Technique	Dilution Refrig?	Funding
NRL,UBC,PNL/USC BU Druikier et al	Granules,Squid	UBC	Operating funds
Stanford Cabrera,Neurhauser Martoff	Ballistic Phonons Tunnel Junctions +trapping	Yes +He^3	Operating funds
UC Santa Barbara Caldwell,Witherell	Quasiparticles?	No low temper.	ββ Decay
UC Berkeley/LBL Sadoulet,Haller,Lange Steiner,Wang,Park	Phonons. Semiconductor Thermistors	Parasiting now 20mK on order	UCB +LBL
Brown Univ. Maris,Seidel,Lanou	Rotons in He^4 Bolometers		Operating funds

The detection of **neutrino coherent scattering** requires also such detectors [6]. This is the main driving force behind the Stanford effort; Cabrera et al. are interested both in detection of reactor and solar neutrinos. Detection of **solar neutrinos** (through neutrino-electron interaction) is the goal of the Brown University effort.

The potential energy resolution and the bulk sensitivity make these detectors very interesting for **double beta decay** [7,6]. UC Santa Barbara is particularly interested in this application.

This energy resolution may be also interesting for electron neutrino mass measurement [8] but as explained by Fiorini [9] at this workshop the problem of pile up is severe.

2. DISCUSSION THEMES

We are therefore witnessing in the USA as in Europe, a burst of interest for cryogenic detectors. However, in the States, most groups are still in the tooling up and exploration phase. It may be interesting to describe some of the discussions going on.

2.1 What is the Best Sensor?

When launching a development, the first decision to take is the kind of sensor to choose, in order to get the low threshold, energy resolution and localization needed. Let us review the candidates.

2.1.1 Superconducting Granules. This method is pushed very actively by Drukier et al. and Kotlicki[10] has described at this workshop their interesting results. However, it is the opinion of this author that granules are too rough bolometers for many applications (including dark matter searches). At low energy they are basically threshold detectors not providing the much needed energy spectrum and the difficulty to make 10^{12} identical detector elements (in order to have 1 kg of sensitive material) may be enormous.

2.1.2 Quasiparticles Detectors. Quasiparticles detection with tunnel junctions [11,12] may provide in particular excellent energy resolution. Caldwell et al. are very much interested in the possible use of this method with ^{100}Mo for double beta decays. An Argonne group (in collaboration with Naples) is developing tunnel junctions and the effort of Cabrera et al. described below (2.1.1) may eventually be applied also to detection of quasiparticles created by particle interactions. Note that bolometry does not work very well with a superconductor since most of the energy is transformed in quasiparticle which have a long recombination time [12].

2.1.3 Phonon Detection through Bolometry.

A first method to detect phonons is basically to measure a temperature rise with a thermistor. This bolometric method is explored at very low temperature with semiconductor thermistors by the Berkeley team. Transition edge superconducting thermistors have not yet attracted any American group [13] because of the very narrow temperature range of operation, the difficulty of coupling with amplifiers because of their low impedance (100Ω) and the variability of film properties. A new bolometric method based on the variation of skin depth in a superconductor has also been proposed [14].

Let us refer finally the reader to the talk of G. Seidel at this workshop [15] for a description of the method proposed by the Brown University team to detect high-energy phonon configuration in liquid helium (rotons) with bolometry.

2.1.4 Phonon Detection with Tunnel Junctions.

A second method to detect phonons of high enough energy (>1 meV) is to use tunnel junctions. Phonons break Cooper pairs in a superconducting film and the created quasiparticle produces a current in a junction. This has been the choice of Cabrera et al. in Stanford. However for this scheme to work, junction areas should be maintained small,and quasiparticles will have to be trappped below the junctions in a way similar to that proposed by Booth [11].

2.1.5 Combination of Phonon and Ionization Detection.

A low-energy recoiling nucleus is providing much less ionization that an electron of the same kinetic energy (for a review see e.g. [16]). The rest of the energy should appear in heat. If both components can be measured, a powerful tool of rejection of interactions on electrons (due to background radioactivity) could be obtained. The Berkeley team is planning to explore this possibility. The main difficulty may be to prevent too large a trapping of ionization products on impurities.

2.2 What is the Importance of Ballistic Phonons?

A second major discussion theme is the role of ballistic phonons in phonon detectors at low temperature.

2.2.1 The Thermal Equilibrium Description .

In the usual description of bolometers [17] it is stated that the intrinsic limit is given by the energy fluctuation in the detector. If it is a crystal of heat capacity C at temperature T, it is given by:

$$\Delta E = \xi \sqrt{(kT^2 C)} = \xi \sqrt{(kT^5 C_o M)} \qquad (1)$$

where ξ is a factor of the order of 2.5 including the Johnson noise of the bolometer. In the second we have used the Debye law:

$$C = C_0 M T^3 \qquad (2)$$

where M is the mass of the crystal. The basic idea [1] of large mass phonon detectors is then to compensate the increase of mass with respect to conventional bolometers by a much smaller decrease of temperature. This description however assumes that phonons are in thermal equilibrium.

2.2.2 Ballistic Phonons

As first noticed by Cabrera in this context [18], this is almost certainly wrong. At very low temperature the mean free path of phonons is very long (i.e. they are ballistic, see e.g. [19]) and the high-energy phonons produced in an interaction will stop cascading down at an energy much higher than the crystal temperature because of the disappearance of anharmonic terms and because of phase space restrictions. The exact energy distribution is poorly known but their average energy is believed to be of the order of 10^{-3} eV.

2.2.3 Impact of Detector Design

Such a property can be helpful in the detector design. It is no more the total heat capacity of the crystal which is relevant but the sensitivity of the sensors. Therefore it may be possible to work at larger temperature or with larger volume. Devices sensitive only to high-energy phonons such as tunnel junctions can be used. Finally, good localisation may be obtained by timing or by focusing (Cabrera [18]).

However in spite of its positive aspects, ballistic propagation of phonons may have detrimental effects (see also [20]) such as a position dependence response due to focusing, a variation from event to event of the non-ballistic component due to e.g. fluctuations in the cascading down or the number of reflections on surfaces defects and isotopic impurities, escape of phonons along some crystal support made of the same material (an effect seen by Moseley et al. in their X ray detector design [21])etc...

Before experimental results are gathered with a variety of sensors, it is difficult to optimize a detector design.

2.3 Parasitic Effects

It should be realized that the extrapolation, say, from the result of MacCammon et al [1c] which use crystals of 10^{-5} grams to 10 grams or 1 kilogram crystal is enormous and requires to work in a much lower temperature region, which is from the point of view of solid state physics, a nearly virgin territory! Let us discuss a few of the questions.

2.3.1 Is the Debye Law Valid at Very Low Temperature?

Note that if the ballistic phonons are indeed predominant the question may be irrelevant.

2.3.2 What is the Effective Phonon Coupling to the Sensors?

Most groups are leaning in the direction of composite detectors where there is a mechanical interface between the sensor and the crystal. Phonon velocity mismatch between the two sides and absorption on one side leads to a Kapitza resistance varying rapidly with temperature. Phonons may then be internally reflected, and not couple effectively with the sensor.

2.3.3 How does the Physics of the Sensors Evolve with Temperature?

For instance the results of Berkeley described below may indicate a decoupling of electrons of the thermistor from the phonons of the lattice, which may prevent proper biasing of the bolometer.

Another example is the problem of leakage current in tunnel junctions which in some cases have a non temperature-dependent component which is poorly understood.

3. EXPERIMENTAL RESULTS

Let us finally turn to the few experimental results which have been gathered so far in the USA. A. Kotlicki [10] described results on superconducting granules and we will not cover them here. We will distinguish between three regions.

3.1 1.4 Kelvin

Such temperatures are easily obtained by pumping on helium and it is possible to have a fast turn around time which makes this region an ideal training ground to learn the basics of the technology and to test electronics.

Detection of α particles is relatively easy and has been achieved at least by three groups : B. Cabrera et al. in Stanford with transition edge thermistors (Fig 1), N. Wang et al. in Berkeley with NTD bolometer (Fig 2) and E. Silver at Livermore. Note in both figures the sharp rise time and the relatively small time constant of the decay.

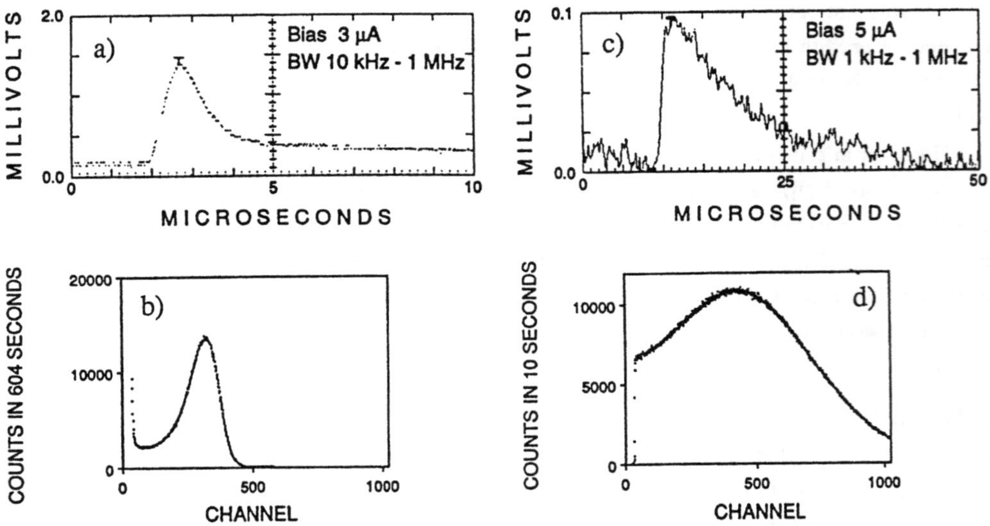

Fig 1. Detection of alpha particles with transition edge detector (from Neuhauser et al [18]).

a) Typical pulse caused by an alpha particle incident on the superconducting film.

b) Corresponding pulse height spectrum

c) Typical pulse when the alpha particle is incident on the back side of the silicon wafer.

d) Corresponding pulse height spectrum.

Usually those results are limited by non-optimal design. For instance the Berkeley result could be theoretically improved by a factor 20 on signal height, a factor 3 by optimal filtering and a factor 10 on the electronics noise and reach eventually the thermal limit of 6kev r.m.s. It remains to be seen whether parasitic phenomena will not limit performance.

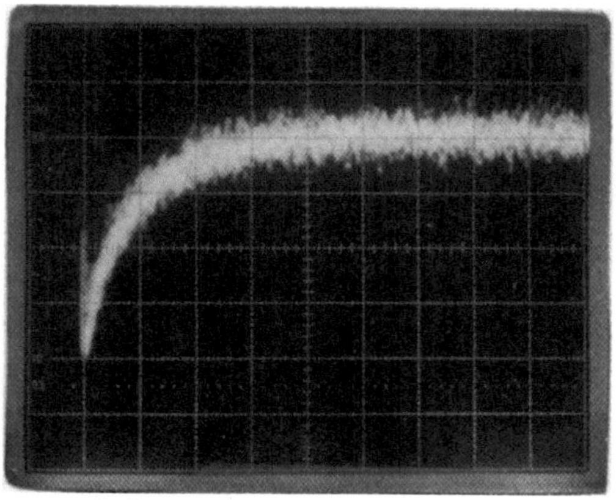

Figure 2. Pulses observed with alpha particle of 5.8 MeV in a bolometer of 6 10^{-3} gram at 1.4K. (N. Wang et al, Berkeley)

3.2 The 300 mk-100mk Region

The Stanford team is tooling up, assembling a 300 mk ^{3}He refrigerator and manufacturing tunnel junctions with lithographic techniques.

The best work for the time being is that of the Wisconsin-Goddard group [1c] working both at 300mk and at 100mk. Their best result so far as a FWHM of 11 eV of the baseline indicating potential threshold of 15 eV for a 10^{-5} gram crystal. However their resolution for 6 keV X rays is degraded presumably by metastable trapping of electrons or holes at impurity sites leading to a fluctuation of the prompt heat release. This effect is minimized by using a composite bolometer with Hg-Cd-Te of approximately zero gap. Their resolution at 6 keV is a record 35eV FWHM. However in the process, the heat capacity has been increased and base line fluctuation have similar width.

This group is now designing their final device with which they hope to go down to few eV FWHM. However,they encounter three problems [21]:

a) Escape of the phonons along the silicon supports of their detector [1c].

b) Parasitic phenomena attributed to phonon electron decoupling or to electric field dependence of the resistance.

c) Excess 1/f noise which does not seem to come from contacts.

3.3 The 20 mK Region

The Berkeley group decided to go directly to low temperature . While waiting for our own dilution refrigerator to be delivered in September, we borrow time on other refrigerators to use Neutron Tansmutation Doped (NTD) germanium thermistors.

Our first attempt at low temperature (confirmed since the workshop by additional measurements) indicated an extreme sensitivity of the thermistor resistance on the electric power dumped into it. 3.10^{-16} Watts are sufficient at 18 mK to induce a significant change of the resistance!

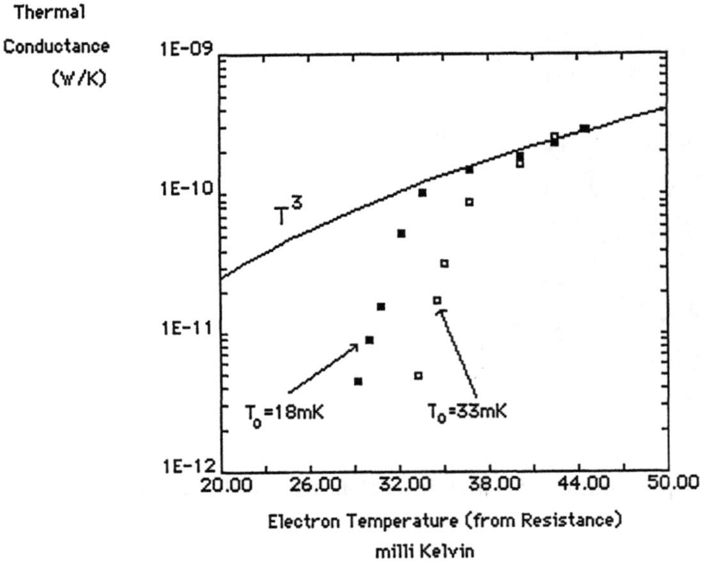

Figure 3. Effective thermal conductance between electron and lattice in a NTD germanium thermistor, as a function of the electron temperature. T_0 is the heat sink temperature. (N. Wang et al, Berkeley).

Because of the weak electric field involved (0.01 V/cm), it is tempting to neglect in first approximation the field dependence of the resistance and compute an effective electron temperature from the measured resistance. The derivative of the electric power with respect to this quantity gives an effective thermal conductance which is plotted in Fig 3 for different bath temperatures. These results are **very preliminary** but are suggestive of a very sharp break-off of the coupling of electrons to the lattice at low temperature. Another explanation could be given in terms of a very strong dependence on the electron field.

We are currently checking in detail this unexpected result and evaluating its impact on the sensitivity of NTD bolometers at very low temperature.

4. CONCLUSION

As shown above, there is a lot of enthusiasm rising in the USA for cryogenic detectors and many groups are tooling up and starting development.

It is clear however, that the questions to solve are as much at the frontier of solid physics as at that of technology, and several years of painstaking effort are ahead before a working device of reasonable size could be made. This will require reasonable funding to escape the common pitfall of superficial, undercritical effort. Coordination and exchange of information will be essential between the various groups in the US and Europe.

References

1. a) Niinikoski, T.O. and Udo , F., Cern Preprint,NP Report 74-6(1974).
   b) Drukier,A.K. and Stodolsky,L., Phys. Rev.,D30, 2295(1984).
   c) Moseley, S.H. and Mather, J.C. and McCammon, D.,J.Appl.Phys.,56(5), 1257(1984).
   d) Cabrera, B. and Krauss, L.M. and Wilczek, F., Phys. Rev.Lett.,55, 25,(1985).

2. The origin of the idea of detecting both ionisation and heat is difficult to trace back. I have heard it mentioned at least by F.Goulding, B.Hyams,T.Niinikoski, and F.Wilczek.
3. Sadoulet,B., in Proc. of the 13th Texas Symposium on Relativistic Astrophysics, Chicago, Dec 1986, in Press
4. Sadoulet,B., Proceedings of the VIIth Moriond Workshop on Searches for New and Exotic Phenomena. Les Arcs Jan 24-31, 1987,in Press.
5. Ahlen,S. et al., Preprint Center for Astrophysics,(1986).
Caldwell,D. et al, in preparation. See also Caldwell in Proceedings of the VIIth Moriond Workshop on Searches for New and Exotic Phenomena. Les Arcs Jan 24-31, 1987,in Press.
6. Cabrera,B. , Caldwell,D. and Sadoulet,B., 1986 Summer Study on the Physics at SSC,Snowmass,(1986).
7. Fiorini, E. and Niinikoski, T.O., Nucl.Instr.and Meth.,224, 83(1984).
8. McCammon,D., Private communication.
9. Fiorini,E., this workshop.
10. Kotlicki,A., this workshop.
11. Booth,N.E., Appl. Phys. Lett. 50 , 293 (1987) and this workshop.
Kraus,H. et al.,Europhys.Lett. 1,161 (1986) and Von Feilitzsch,F., this workshop.
12. Twerenbold,D.,Europhys. Lett. 1,209 (1986).
13. See,however, the description of the Munich effort in this workshop.
14. Mac Donald,D.G.,"Novel Superconducting Thermometer for Bolometric Applications", National Bureau of Standards (Boulder) preprint, 1987.
15. Seidel,G., this workshop.
16. Sadoulet,B. et al "Testing the WIMP explanation of the solar neutrino puzzle with conventional silicon detectors". Lawrence Berkeley Laboratory preprint April 87.
17. Mather, J.C., Applied Optics,21, 1125(1982), 23, 584(1984), 23, 3181(1984).
18. Cabrera,B.,Martoff,C.J.,and Neuhauser,B., "Acoustic Detection of Single Particles", submitted to Mucl. Instr. and Methods.
Neuhauser,B., Cabrera,B., Martoff,C.J. and Young,B.A., "Acoustic Detection of Single Particles for Neutrino Experiments and Dark Matter Searches". , 1986 Applied Superconductivity Conference,Stanford University preprint BC 50-86.
19. Bron,W.E, ed.,"Non equilibrium Phonon Dynamics", NATO ASI Series B124, Plenum Press, New York, 1985.
20. Maris,H.J.,Fifth International Conference on Phonon Scattering in Condensed Matter, Urbana, Illinois, June 2-6, 1986.
21. Moseley,H., McCammon,D., private communication.

Calorimetric Detectors at Low Temperatures

F.v. Feilitzsch, T. Hertrich, H. Kraus, L. Oberauer, Th. Peterreins,
F. Pröbst, and **W. Seidel**
Physik-Department, Technische Universität München,
D-8046 Garching, Fed. Rep. of Germany

Properties of calorimetric detectors at low temperatures are discussed. Metastable superconductors, superconductors at the critical temperature used as thermometers and superconducting tunnel junctions are considered as sensors for energy absorption. The different mechanisms of detection are compared and eventually achievable sensitivities estimated.

1. Introduction

In the past various calorimetric detection methods using low temperatures to increase the sensitivity were suggested and first successful developments gave promising results. This led to a number of ambitious proposals for the application of this new type of detectors, among others X-ray, neutrino, dark matter, and magnetic monopole detection [1-6].

For a successful application of these low temperature detectors a detailed understanding of the solid state properties at the temperatures under discussion is essential. We report on experimental tests of the phase transitions of metastable superconducting grains (section 2), superconducting phase-transition thermometers (section 3) and superconducting tunnel junctions (section 4). In addition we give a short review of the properties of some possible low temperature detectors.

2. Metastable Superconducting Grains [7]

2.1 Test of Sn-Grains

Tests were performed with Sn grains of diameters 5 $\mu m < d < 25$ μm which were embedded in wax and surrounded by a read-out loop of 1 cm length and 100-200 μm width (fig. 1). Perpendicular to the loop a magnetic field was applied. The phase transitions of individual grains were detected via a change of the magnetic flux through the read-out loop due to the Meißner effect. Even for low concentrations of grains in the wax, the spread ΔH of the phase transitions was observed to be of the order of $\Delta H/H_{sh} \approx 30\%$ independent of the temperature

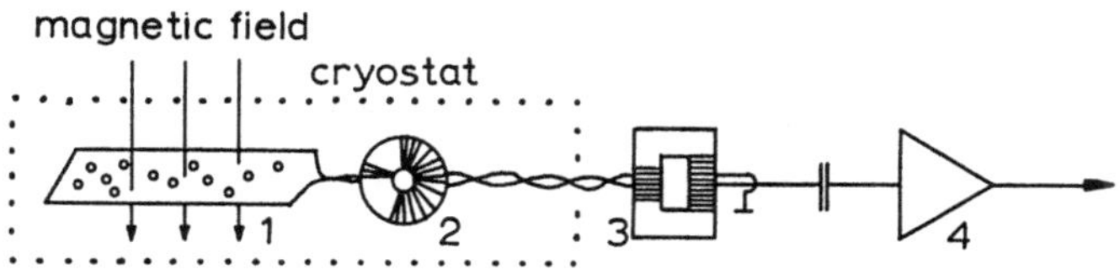

Fig. 1. Schematic diagram of the readout system with pickup loop (1), two transformers (2,3) and preamplifier (4).

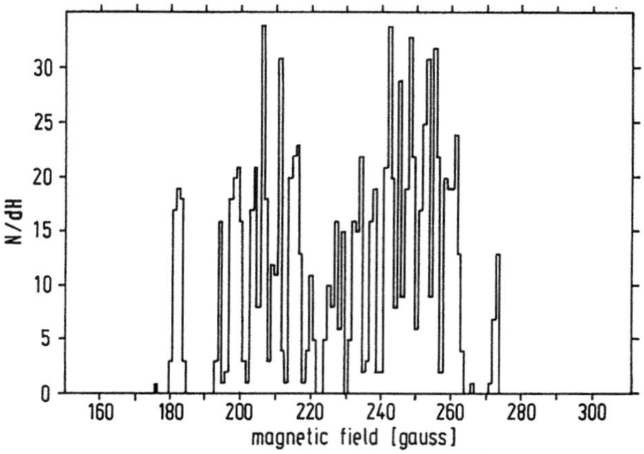

Fig. 2. Superheating phase transitions of 52 Sn-grains in one loop as a function of applied magnetic field. 20 runs were added up (dH = 1.0 Gauss).

$T < T_c$,were H_{sh} is the superheating field and T_c the critical temperature. In fig. 2 the phase transitions of 52 Sn grains from the superconducting to the normal conducting state are added up for 20 upward magnetic field sweeps between 0 Gauss and 300 Gauss. This shows that within the experimental resolution each individual grain undergoes a phase transition in a reproducible way ($\delta H/H \approx 0.5\%$) whereas the whole sample shows a spread of $\Delta H/H = 30\%$. From the low filling factor used in the experiment one may conclude that this spread results from varying individual properties of the grains rather than from diamagnetic perturbations of the applied magnetic field.

This large ΔH leads to the very low sensitivity ϵ_γ of the sample of grains for γ-ray detection which was measured to be $\epsilon_\gamma \leq 0.3\%$ for 14 keV γ-rays. Here ϵ_γ denotes the percentage of grains being sensitive for detection, i.e. which are close enough at the phase boundary to undergo a γ-induced phase transition.

2.2 Tests of Cd-Grains

As the required drastic improvement of the sensitivity of the grains seems presently not to be achievable by improved grain production techniques an investigation of lower T_c superconductors is suggested. Tests were performed with Cd which has a critical temperature of T_c=540 mK. Due to the lower T_c and the correspondingly lower critical magnetic fields a signal detection as used for the Sn grains would have required a SQUID read-out system. Therefore a different technique was used. About $3.8 \cdot 10^6$ grains embedded in Apiezon N grease were filled in an inductance coil of 10mm height and 3mm inner diameter, which corresponds to a filling factor of 3.8% in volume. The inductance of the coil, depending on the number of grains in the superconducting state, was measured via an L-C resonant circuit. The stability of the resonance frequency was better than 10^{-6} during five minutes.

The experiments with Cd-grains were performed in a dilution refrigerator at temperatures ranging from 50 mK to 400 mK. Temperatures were measured with a calibrated germanium resistor and two carbon resistors.

2.2.1 Hysteresis Curves

Figure 3 shows a hysteresis curve which was obtained by measuring the frequency while sweeping the magnetic field and keeping the temperature at T=350 mK. The highest frequency is observed when all grains are superconducting, the lowest when all grains are normalconducting. The spread of the phase boundaries was $\Delta H/H_{sh} \approx 14\%$.

At T $\leq$ 350 mK an instability of the transitions from the superconducting to the normalconducting phase was observed. A large fraction of grains changed the phase simultaneously at the same value of the applied field independent of the speed of the magnetic field variation. Figure 4 shows the effect at temperatures T = 300 mK, 200 mK and 135 mK. This observation

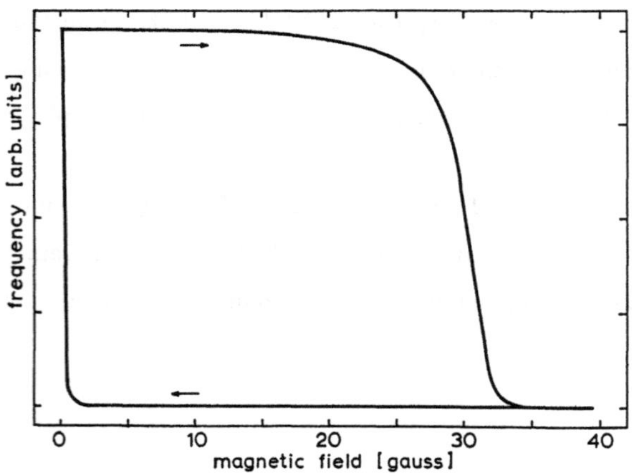

Fig. 3. Hysteresis curve of Cd at T=350mK. The resonance frequency of the oscillator indicating the amount of superconducting grains in the sample is shown as a function of the applied magnetic field.

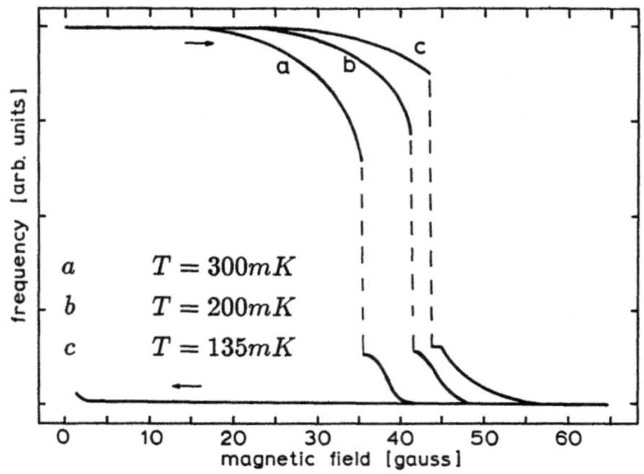

Fig. 4. Hysteresis curves showing an avalanche effect measured at different temperatures.

may be tentatively explained by the release of latent heat at the phase transition of a superheated grain. If this heat is large enough to flip at least one neighbouring grain a chain reaction takes place. A more detailed discussion is given in Ref. [7]. This effect has to be taken into account for the design of a detector based on a large sample of superconducting grains.

2.2.2 Irradiation with a γ-Source

The sample of Cd grains was irradiated with a source containing ^{67}Ga (E_γ=93 keV, $T_{1/2}$=78 h) and ^{65}Zn (E_γ=1.1 MeV, $T_{1/2}$=244 d). With the time dependence of the ^{67}Ga activity a systematic test of the detection technique was performed. The experiment was made at temperatures close to T_c where no instability of the phase transition was observed. The change of the resonance frequency, representing the number of grains flipped due to irradiation, was used to test the sensitivity of the grains. At the beginning of the irradiation on the average 3.2 grains were flipped by one γ-quant at a temperature of T=414 mK. More than 90% of the grains were flipped after a long time of irradiation.

Using a three dimensional Monte-Carlo calculation an energy threshold for the grains was estimated by simulating the energy loss of photo and compton electrons in the sample. The measured data were well reproduced for an energy threshold of $Q/V = 12eV/\mu m^3$ where Q is the deposited energy and V the volume of the grains.

The observed sensitivity of Cd grains which is drastically higher than that of Sn grains can be attributed mainly to the lower critical temperature of Cd. Therefore calculations of the sensitivity as a function of operating temperature were performed for superconductors with even lower T_c, namely W (T_c=12 mK) and Be (T_c=25 mK). In this calculation the entire grain is considered to be heated up homogeneously before it undergoes a phase transition. Since the shape of the phase boundary is well known every change in temperature ΔT can be translated into a change of magnetic field δH. A certain energy deposition per unit volume Q/V leads to a rise in temperature which corresponds to a change in magnetic field. The relative change in the magnetic field $\delta H/H_a$ (H_a: applied magnetic field) for a given energy deposition Q/V calculated for W is shown in fig. 5. For a fully sensitive sample the relative change in magnetic field due to an energy deposition Q/V should be bigger than the observed spread in the hysteresis curve (for our Cd grains $\delta H/H > 14\%$).

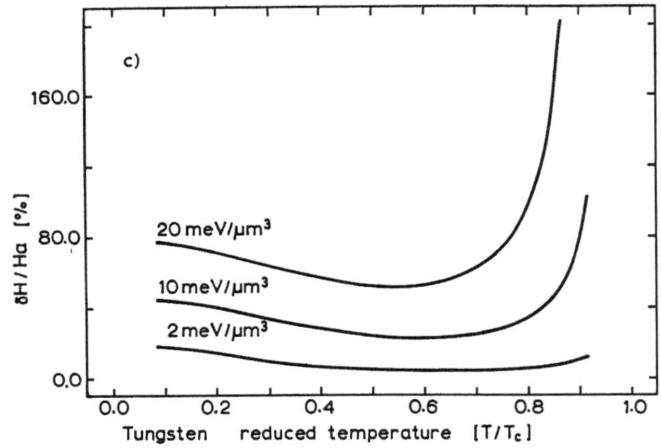

Fig. 5. Calculated sensitivities of tungsten. The x-axis shows the reduced temperature T/T_c at which the system is operated. The assumed energy deposition per μm^3 is indicated at the left side of the curves. Such an energy deposition leads to a temperature change ΔT of the grain, which corresponds to a change in magnetic field δH. The relative change $\delta H/H_a$ is shown at the y-axis. H_a is the applied magnetic field.

3. Composite Calorimeters

A composite calorimetric detector consists of an absorber, serving for the absorption of radiation, and a separate thermometer to detect the corresponding change of temperature. An evident advantage of this concept is the possibility of independent optimization of the properties of both components. For such systems a variety of realizations has been suggested and tested in the past [1,8,9,10].

A calorimeter of this type is characterized by the following parameters:

heat capacity $$C = (c_{lattice} + c_{electrons}) \cdot M[\tfrac{J}{K}]$$
$$M = \text{mass of absorber}$$

coupling to a heat sink	$g\ [\frac{W}{K}]$
responsivity of calorimeter	$R_v = \frac{\Delta U}{P}\ [\frac{V}{W}]$
given as output voltage per incident power	
el. resistance of thermometer	$R[\Omega]$
radiative constant	$\sigma_E\ [\frac{W}{cm^2 K^4}]$
background temperature	$T_B[K]$
surface of calorimeter	$A[cm^2]$
bias current	$I_B[A]$

The energy resolution of the system is limited by the noise, which is produced by a variety of uncorrelated noise sources. If Johnson noise is the dominant electronic contribution, the total noise equivalent power [NEP] for signal detection is given by [11]:

$$NEP^2 = \underbrace{\frac{4kTR}{R_v^2}}_{\text{Johnson noise}} + \underbrace{4kT^2 g}_{\text{thermal noise}} + \underbrace{8\sigma_E k T_B^5 A}_{\text{background fluctuations}}$$

It can be seen that lowering the operating temperature reduces all contributions to the total NEP and that a high responsivity of the thermometer is needed.

3.1 Absorber

The specific heat of ideal nonmagnetic cubic solids is determined from contributions of lattice vibrations C_ℓ and conduction electrons C_e:

$$C = C_\ell + C_e$$

For materials following the Debye law $C_\ell \sim (T/\Theta_D)^3$, where Θ_D is the Debye temperature. The specific heat contribution from conduction electrons C_e is given by:

$C_e = 0$	for dielectric solids
$C_e = \alpha \cdot T$	for metals
$C_e = b \cdot \exp(-\Delta/kT)$	for superconductors at $T << T_c$

Δ is the energy gap of the superconductor, k the Boltzmann constant and b a normalisation constant. It is evident, that insulators with a high Debye temperature Θ and superconductors for $T/T_c \ll 1$ will have a very low specific heat at low temperatures. This leads to a correspondingly high change in temperature for a given absorbed energy. There are however several problems which have to be taken into account.

a) The absorber material has to be homogeneously cooled down to the operating temperature. This can be difficult if big masses are required and the heat conductance becomes poor at low temperatures.

b) Not all of the absorbed energy may be converted into a temperature rise of the calorimeter. Some energy might get lost by light emission or be stored in long living excitations in the absorber. The energy resolution of the system is affected by statistical fluctuations of the amount of energy not converted into heat.

The limit set on the energy resolution by the creation of Frenkel pairs (vacancy interstitial pair) in diamond has been estimated [12] to be

$$\Delta E(FWHM) \quad = \quad 2.35 Q_s \sqrt{N}$$

where Q_s is the binding energy of a Frenkel pair and N the number of created pairs. For diamond Q_s is approximately 12eV. The number of actually created Frenkel pairs strongly depends on the mass and energy of the interacting particle. Electrons create no Frenkel pairs therefore ΔE_e=0. From a 5.5 MeV α-particle about 92 Frenkel pairs are created, leading to $\Delta E_\alpha \approx$270 eV [12]. Similar effects are expected from metastable electronic states which however are also produced by γ-ray and electron irradiation. They should not exist in metals.

The luminescence efficiency in diamond is of the order of 1% and is expected to limit the energy resolution to $\Delta E_{lum} \approx$ 1 keV [12] for 5.5 MeV α-particles, if the light cannot be reabsorbed within the crystal.

3.2 Thermometers

A variety of thermometers has been considered and investigated to serve for the read out of the rise in temperature following a radiation absorption in a calorimetric low temperature detector. These are, among others, semiconducting thermistors [1,8,9,10], paramagnetic salts [8], superconducting transition edge thermometers. Impressive results were obtained from semiconducting thermometers on a silicon absorber for X-ray detection leading to an energy resolution of $\Delta E \approx 30eV$ for 6 keV X-rays from a ^{55}Fe source [1].

The sharp resistance change at the superconducting phase transition can be used for thermometry in a small temperature range. Via an applied external magnetic field the phase transition can be broadened to increase the dynamic range of the thermometer at the expense of a reduced responsivity. These transition edge thermometers are known as a powerful tool in solid state physics to study phonon propagation with high time resolution [13].

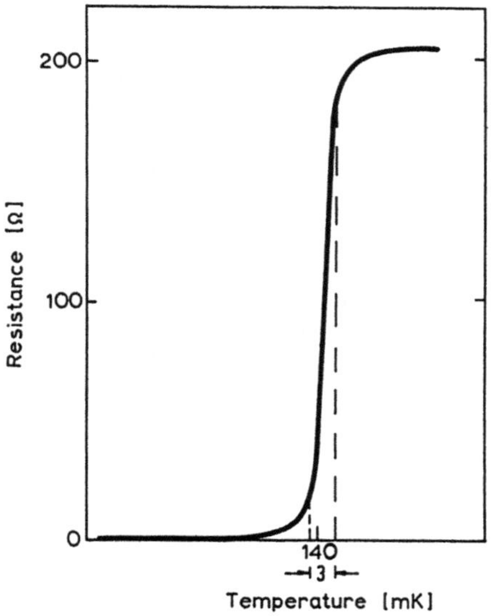

Fig. 6. Transition curve of an Ir-Film of dimensions 0.1 μm × 100 μm × 1000 μm . Measuring current 1μA.

We investigated an Iridium transition edge thermometer. A transition curve from an Ir-film of dimensions 0.1 μm × 100 μm × 1000 μm. is shown in fig. 6. The phase transition occurred at T = 140 mK, with a normal conducting resistivity of R_{nc}=200 Ω. The transition from 13% to 87% of R_{nc} was essentially linear in temperature and occurred within ΔT=3 mK. A responsivity of $\Delta R/\Delta T \approx 5 \cdot 10^4\ \Omega K^{-1}$ was achieved. A normal conducting specific heat of $C_{Ir} \approx 5 \cdot 10^{-13}$ J/K of the Ir thermometer at T=140 mK is estimated.

Due to the low electric resistance of the thermometer the total noise is dominated by the amplifier noise contribution. Using a bias current of I_B=1 μA and a noise level of $\Delta U_{noise} = 10^{-7}$ V for the voltage measurement we can estimate the sensitivity ΔQ of the device:

$$\Delta U = \Delta T \frac{dR}{dT} \cdot I_B$$

$$\Delta T = \frac{\Delta Q}{C_{Ir}}$$

$$\Delta Q = \frac{\Delta U \cdot C_{Ir}}{\frac{dR}{dt} \cdot I_B}$$

$$\Delta Q_{noise} \approx 10^{-18} Joule$$

$$\approx 6eV$$

The noise level of the voltage measurement would be considerably improved, if a SQUID would be used.

We also investigated Al thin film transition edge thermometers. The advantage of these thermometers is an operating temperature slightly above 1 K which can be reached with a ^{4}He cryostat. The films were produced by vacuum evaporation. The rest gas composition in the vacuum system was found to determine the quality of the films. To get films with a narrow transition, i.e. a high responsivity, the evaporation had to be performed at a total pressure of $P < 10^{-7}$ mbar. The oxygen partial pressure was $P_{O_2} < 5 \cdot 10^{-8}$ mbar.

Al films were found to change their properties when exposed to air. To protect the films against oxidation a layer of 400 Å Mg F_2 was evaporated on top of them. In fig. 7 transition curves of an Al film obtained after evaporation, at the second cycle to $1.3K$ and after 16h of exposure to air at room temperature are shown for comparison. It can be seen, that thermal cycling and exposing the films to air does not affect the the transition curve. The curves of our best stripes represent a responsivity of $\Delta R/\Delta T$=325 Ω/K.

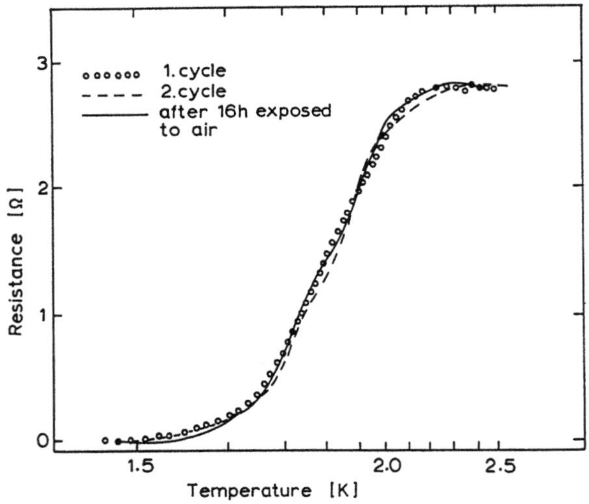

Fig. 7. Superconducting transition of an Al-film, protected against oxidation with a 400 Å Mg F_2-film on top of it. Open circles represent data from the first cooling cycle. The dashed curve was obtained in the second cycle after warming up to room temperature in He-atmosphere. The fully drawn curve was taken in a third cycle after 16 h of exposure to air.

4. High-Resolution X-Ray Detection with Superconducting Tunnel Junctions

The potential high energy resolving power of superconducting detectors arises from the small energy gap Δ in the order of meV in the electronic excitation spectrum of superconductors, which is about three orders of magnitude smaller than in semiconductors. In contrast to semiconducting detectors, energy spent in excited phonons is not necessarily lost. As their energies can be considerably larger than the energy gap, they can break Cooper pairs and contribute to the signal. Consequently, the statistical limitation on the resolution may be lower by more than one order of magnitude as compared to semiconducting detectors.

4.1 Basic Principles

4.1.1 Quasiparticle Tunneling

It is well known that between normal metals separated by a thin barrier (thickness ≈ 20Å) quantum mechanical tunneling of electrons is possible. When the metals become superconducting the free electrons near the Fermi level (E_F) are bound in Cooper pairs. These pairs are occasionally broken into two quasiparticles which in turn can recombine to emit a phonon. The resulting equilibrium density of quasiparticles at temperature T is approximately proportional to the Boltzmann factor $\exp(-\Delta/kT)$. Tunneling of electrons through the barrier mediates tunneling processes of quasiparticles. This can be understood easily by expressing the electron operators in the tunneling Hamiltonian by quasiparticle operators [14]. For bias voltages U in the range $kT/e \ll U < 2\Delta/e$ only the two tunneling channels shown in fig. 8 contribute to the current flow. Quasiparticles from the left can tunnel into empty states of the right superconductor (process a). In process b of fig. 8 a pair on the left side is broken into two quasiparticles, one of them combines with a quasiparticle to form a pair on the right side of the barrier, the other goes into empty states of the left superconductor. Both processes shown in fig. 8 involve the flow of one electronic charge from left to the right, but in process b the quasiparticle is tunneling in the opposite direction. A quasiparticle can therefore, if it does not recombine or get lost otherwise, tunnel back and forth, each time contributing the flow of one electronic

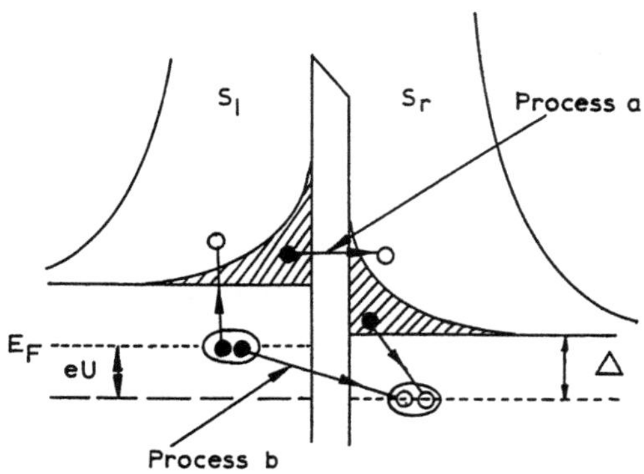

Fig. 8. Quasiparticle tunneling between two superconductors S_ℓ, S_r for $0 < U < 2\Delta/e$. In process 1 a a quasiparticle tunnels from an occupied state in the left superconductor S_ℓ into an empty state of the superconductor S_r on the right side of the barrier. In process b a Cooper pair is broken to create one quasiparticle in an empty state of S_ℓ. The second quasiparticle tunnels and combines with another to form a pair in S_r.

charge from the left to the right. The tunneling current $I_{a,b}$ of the two channels, assuming two identical superconductors, is given by:

$$I_a = e\gamma_a N_\ell \tag{1a}$$

$$I_b = e\gamma_b N_r \tag{1b}$$

$$\gamma_a = \frac{1}{e^2 R_{nn} A \cdot 4 \cdot n(0)} \cdot \frac{\Delta + eU}{\sqrt{(\Delta + eU)^2 - \Delta^2}} \cdot \frac{1}{d_\ell} \tag{2a}$$

$$\gamma_b = \frac{1}{e^2 R_{nn} A \cdot 4 \cdot n(0)} \cdot \frac{\Delta + eU}{\sqrt{(\Delta + eU)^2 - \Delta^2}} \cdot \frac{1}{d_r} \tag{2b}$$

N_ℓ, N_r is the number of quasiparticles on the left and right side of the barrier. The tunneling probabilities γ_a, γ_b are inversely proportional to the thicknesses d_ℓ, d_r of the left and right superconducting film. R_{nn} is the normal conducting resistance of the junction, $n(0)$ is the single-spin density of states at the Fermi surface and A is the junction area. In thermal equilibrium both channels equally contribute to the electrical tunnel current, whereas the net quasiparticle flow cancels. An extra tunnel current resulting from the excess number of quasiparticles produced by absorption of ionizing radiation in one of the films is also given by (1a,b).

4.1.2 Relaxation Processes

When ionizing radiation deposits energy in one of the films of the junction, highly excited quasiparticles are created. Within a very short time of the order of 10^{-10} s these quasiparticles relax to the gap edge by further breaking of Cooper pairs and emission of relaxation phonons. About 60% [15] of the deposited energy is 'stored' in quasiparticle excitations at the gap edge, the rest is spent in phonons of energy lower than 2Δ. The onsetting tunneling has to compete now with two loss processes, namely the recombination under continuous emission and reabsorption of 2Δ-phonons and the diffusion of quasiparticles out of the tunneling zone. Dependent on the relative strength of these processes a noticeable quasiparticle density may eventually build up in the film not hit by the X-ray, allowing each quasiparticle to tunnel several times. Because of their importance these loss processes will now be considered in some detail.

For higher temperatures, when the excess density of quasiparticles is small compared to their thermal density, the recombination probability γ_R follows the temperature dependence of the thermal quasiparticle density which is approximately proportional to the Boltzmann factor $\exp(-\Delta/kT)$. At $T \approx 700$ mK the thermal recombination rate equals the tunneling rate for our tin junctions. At still lower temperatures recombination is determined by the actual excess density. In the junctions used in our experiments (fig.9b) there are roughly 10 thermally excited quasiparticles at T=0.3 K, whereas a 6 keV X-ray produces about $6 \cdot 10^6$ quasiparticles.

The mean free path l of quasiparticles is determined by elastic scattering at impurities which can be calculated from film resistivity above T_c. The distance L which a quasiparticle diffuses within time t is found from random walk (v_F: Fermi velocity):

$$L = \sqrt{\frac{v_F \cdot l \cdot t}{3}} \tag{3}$$

A detailed quantitative understanding of the quasiparticle dynamics may be gained by solving a set of coupled rate equations for quasiparticles and 2Δ-phonons in both films. These equations are an extension of the well known Rothwarf-Taylor equations [16]. Since a complete analysis of them would be beyond the scope of this article, only the basic processes that have to be included will be briefly mentioned:

1. tunneling of quasiparticles from one film into the other and vice versa
2. diffusion of quasiparticles and phonons within the films
3. recombination of quasiparticles by emission of 2Δ-phonons and pair breaking by reabsorption of these phonons
4. exchange of phonons between the films and their escape into the substrate.

The rate constants for process 1 (see (2a,b)) may be determined from normal tunneling resistance R_{nn} and the film thicknesses, the diffusion constant may be extracted from film resistivity above T_c, recombination and pair breaking as well as phonon escape time constants are reasonably well known from literature [17,18]. Assuming a transmission probability of one for the transmission of phonons through the thin oxide layer separating the superconducting films, all the basic rate constants are determined and the set of equations should adequately describe the response of the detector.

4.2 Experimental Details

Sn tunnel junctions were produced by vacuum evaporation of Sn through suitable masks and barrier formation by glow discharge oxidation. Two different junction geometries shown in fig. 9a,b were used in the irradiation experiments, with junction areas, defined by the overlap of the two films, of about $8.8 \cdot 10^{-2}$ mm^2 and $1 \cdot 10^{-2}$ mm^2, respectively.

The samples were mounted in a vacuum chamber inside a ^{3}He-cryostat. Temperature was measured with a Ge-resistance thermometer attached onto the cold finger close to the junction. A weak magnetic field (20 G - 100 G) was applied parallel to the surface of the junction to suppress the d.c. Josephson current.

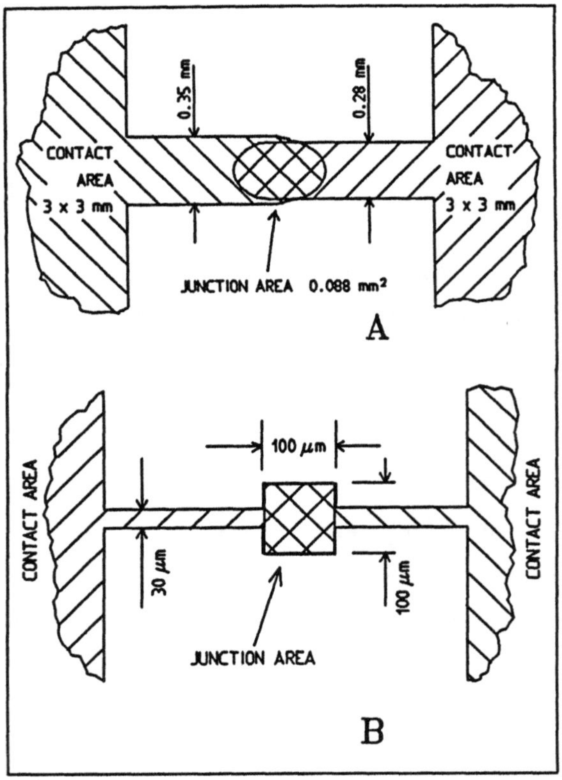

Fig. 9. Geometries of the detector. The tunnel junction is confined to the overlap of the two superconducting Sn films (doubly shaded area).

The detector was illuminated by a ^{55}Fe radioactive source, emitting K_α, K_β X-rays at 5.89 keV (88%) and 6.49 keV (12%), respectively. In order to adjust the X-ray flux and to absorb Auger electrons emitted by the source, a mylar foil was inserted between source and detector. A typical count rate in the experiments was 250 Hz.

A d.c. bias was supplied with a current source. The junction signals were transmitted along a 50Ω coaxial cable to a charge sensitive preamplifier operated at room temperature. Pulse height spectra were recorded with a multichannel analyzer after suitable pulse shaping ($3\mu s$ to $10\mu s$).

4.3 Results

4.3.1 Results from Junctions of Simple Geometry[19]

The first irradiation experiments were performed with junctions of the simple geometry shown in fig. 9a. A current-voltage characteristic, recorded at 0.3 K, is displayed in fig. 10. It was taken

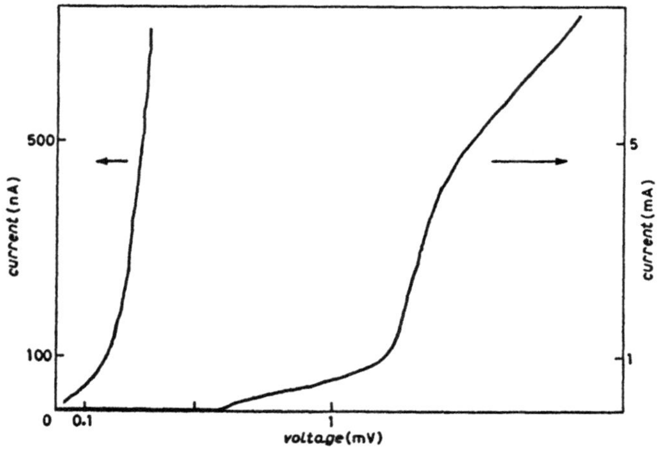

Fig. 10. Current-voltage characteristic at T=0.3 K of a $Sn/SnO_x/Sn$ superconducting tunnel junction used for X-ray detection. Two different current scales are given. The scale on the right applies to the right curve and the scale on the left to the left curve. The current rise at $U=\Delta/e=0.6$ mV is probably due to two-quasiparticle tunneling . Typical bias currents for detector operation lie in the range from 20 nA to 50 nA.

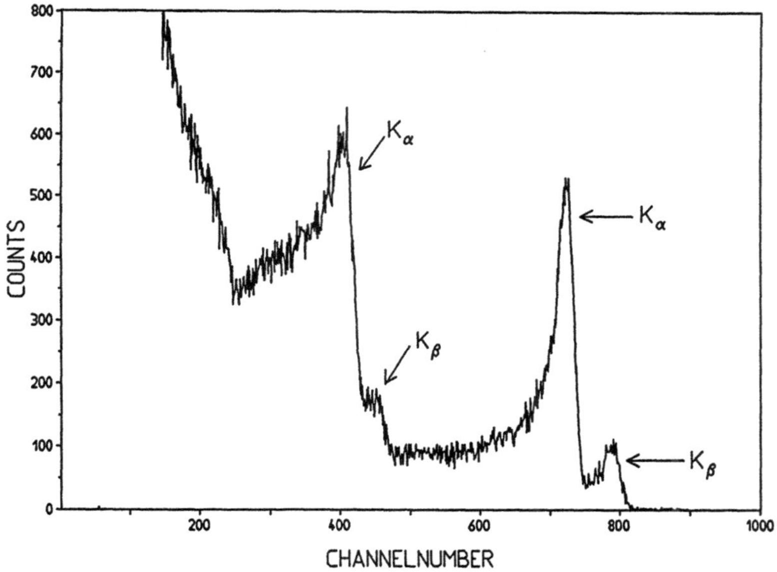

Fig. 11. Pulse height spectrum recorded at T=0.3 K with a superconducting $Sn/SnO_x/Sn$ tunnel junction of the geometry shown in fig. 9a, exposed to a ^{55}Fe source emitting Mn K_α, K_β X-rays with energies of 5.89 keV and 6.49 keV, respectively. The upper pair of peaks is due to X-ray absorption in the 260 nm thick top film and the lower pair of peaks corresponds to detection in the 390 nm thick bottom film. A resolution of $\Delta E=250$ eV (FWHM) can be inferred from the response of the top film.

from the junction which was used to record the spectrum shown in fig. 11. The displayed curve shows some deviation from the ideal single-particle tunneling pattern, namely a pronounced current increase at $U=\Delta/e$, i.e. at U=0.6 mV for Sn. At voltages less than Δ the current decreased strongly with temperature down to 0.3 K, whereas above $U=\Delta/e$ the current remained at a constant value below 1 K. This behaviour may be attributed to a significant contribution of two quasiparticle tunneling, which is expected to appear at bias voltages exceeding Δ/e. The ratio of R_D/R_{nn}, with R_D being the differential resistance at the bias point, was found a good measure of the signal to electronic noise ratio obtained in the measurements. For $R_D/R_{nn} \geq 10^3$ pulses emerge out of the noise, for $R_D/R_{nn} \geq 10^5$ resolution is no more dominated by electronic noise.

A representative pulse height spectrum recorded at 0.3 K is shown in fig. 11. It results from irradiating a junction with a top film of 260 nm and a bottom film of 380 nm thickness. In the upper part of the spectrum the well resolved pair of peaks corresponds to Mn K_α and K_β photons which were absorbed in the top film. From the observed peak width an energy resolution of ΔE=250 eV (FWHM) is derived. The lower pair of peaks results from K_α, K_β X-rays absorbed in the bottom film.

This interpretation was confirmed by a series of experiments with junctions of varying film thicknesses. This systematic study reveals that the pulse height is inversely proportional to the thickness of the corresponding film. This can be seen in fig. 12, where the ratio of pulse heights

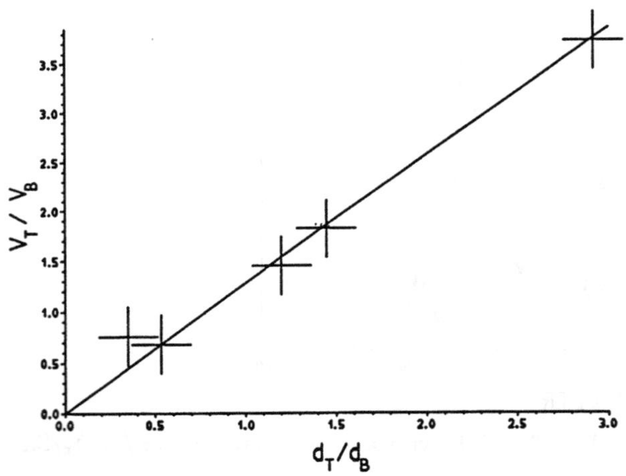

Fig. 12. Dependence of the ratio of pulse heights, resulting from absorption of Mn K_α X-rays in the two films of $Sn/SnO_x/Sn$ tunnel junctions, on the inverse ratio of the corresponding film thicknesses. V_T, V_B are the pulse heights from signals originating from top and bottom film and d_T, d_B denote the thicknesses of top and bottom film, respectively.

obtained for the K_α peaks from the two films of a junction were plotted versus the inverse ratio of the corresponding film thicknesses. From this inverse proportionality one may conclude that quasiparticles only tunnel once. Loss processes, presumably diffusion losses, which are fast compared to tunneling prevent the formation of any significant density of quasiparticles in the film which was not hit by the X-ray. The slope of the straight line plotted in fig. 12 is s=1.27, which means that signals from the bottom film (V_B in fig. 12), which is in contact with the substrate, are always smaller in amplitude compared to signals from top films (V_T) of similar thicknesses. The nature of this additional loss process is not understood. Appreciable phonon losses at the junction-substrate interface most likely do not explain this large effect, since solutions of the rate equations with any reasonable phonon transmission probability through the film-substrate boundary failed to reproduce such a large effect. Moreover no striking difference between single crystalline Si and SiO_2 glass substrates was observed in the experiments. Scanning electron microscope pictures revealed that the top film which is deposited after glow discharge oxidation of the base film always exhibited substantial higher granularity than the first deposited film. The higher granularity might slow down the diffusion of quasiparticles in the top film, thus leading, via a reduction of diffusion losses, to larger pulses.

The width and asymmetry of the peaks may be attributed to quasiparticle diffusion. Depending on the position of the X-ray absorption within the tunneling zone, varying fractions of quasiparticles are lost for the tunneling process by diffusing into the contact leads. The almost constant background towards lower energies results from quasiparticle diffusion from the leads into the junction volume. From the measured conductivity of $\sigma = 4.6 \cdot 10^5 (\Omega cm)^{-1}$ of the tin films at 4.2 K a mean free path for quasiparticles of l=2100 Å was estimated . Using (3), a distance of L=1.6 mm which a quasiparticle travels within the 8 μ s rise time of the pulses was calculated. Since the source illuminated homogeneously a region of 2 mm in diameter around the junction, varying fractions of quasi-particles generated by X-rays hitting the Sn-films to the left and right of the junction (fig. 9a) may have diffused into the tunneling volume, thereby creating pulses of reduced amplitude. The steep rise in count rate at low amplitudes (left to the lower pair of peaks in the spectrum of fig. 11) is not an effect of electronic noise. Numerical solutions of the rate equations revealed that this background may also originate from quasiparticle diffusion from the leads into the junction, but X-ray absorption in the substrate with subsequent phonon detection in the junction volume may also contribute to such a background.

From a comparison of the X-ray induced pulse height with the one obtained from putting a test charge on a small capacitor at the input of the preamplifier the collected charge was estimated. In several experiments with junctions of this simple geometry the number of collected charge carriers was limited to 30% of the primarily produced quasiparticles. These estimations were based on an effective amount of energy Ω needed for the creation of a single quasiparticle of $\Omega = 1.7\Delta$, which is calculated in Ref. [15].

4.3.2 Results from Junctions of Improved Geometry

The pronounced broadening of the left side of the peaks, the constant background at the lower energy side of the peaks, the poor charge collection as well as the large distance L which quasiparticles travel within the rise time of the pulses already indicate that diffusion is an important loss process with strong influence on spectral shape and resolution. To study the diffusion in more detail experiments with junctions of the improved geometry shown in fig. 9b were performed. A pulse height spectrum taken at 0.3 K is shown in fig. 13. The broadening of the left side of the peaks almost disappeared. The background to the left of the upper pair of peaks is also strongly reduced, consistent with the smaller dimensions of the contact leads. Despite the reduction of diffusion losses the resolution, extracted from the upper pair of peaks in fig. 13, only slightly improved. This can be understood from the contribution of electronic noise which strongly deteriorated the resolution obtained from this junction, whereas the peak width of the spectrum in fig. 11, which was taken from a junction with unbounded diffusion,

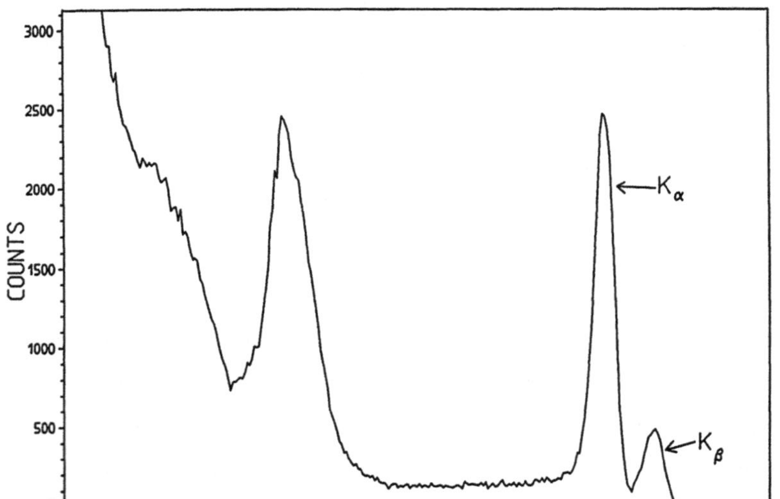

Fig. 13. Pulse height spectrum recorded at T=0.3 K with a superconducting $Sn/SnO_x/Sn$ tunnel junction of the geometry shown in fig. 9b, exposed to a ^{55}Fe source. The upper pair of peaks is due to X-ray absorption in the 450 nm thick top film, the lower peaks correspond to detection in the 505 nm thick bottom film.

had only a minor contribution from electronic noise. The increased electronic noise can be attributed to the minor quality of this junction characterized by $R_D/R_{nn} \approx 2.5 \cdot 10^3$, whereas the junction yielding the spectrum in fig. 11 gave $R_D/R_{nn} \approx 1.3 \cdot 10^4$. With junctions of exactly the geometry of fig. 9b but considerably better quality a resolution of 67 eV has been obtained [20,21].

To find out to what extent quasiparticle diffusion still affects the spectra obtained from junctions of the improved geometry (fig. 9b), the above mentioned rate equations were solved numerically. In a first calculation junctions of the simple geometry (fig. 9a) on a silicon substrate were investigated. The solutions reproduced well the experimentally observed magnitude and time dependence of the excess tunneling current, giving us confidence that the dynamics of quasiparticles is reasonably described by the simple rate equations. The decay time of the current of several μs is entirely determined by the diffusion of quasiparticles into the current leads. A second solution was obtained for a junction of the same geometry but without any diffusion losses into the contact leads. The only way to reach thermal equilibrium in this case is the escape of the recombination phonons into the substrate. The decay times of the excess tunneling current increased from some μs with diffusion losses to several ms for the junction with diffusion losses switched off. The collected charge obtained by integrating the excess current also increased by about a factor of 1000. Preventing the quasiparticles to diffuse into the contact leads thus would largely improve the signal to electronic noise ratio. The undesirable doubling of the peaks should also disappear, since after a short transient time of the order of the tunneling time the quasiparticle density is the same in both films, independent of which film has been hit by the X-ray.

Experimentally the trapping of quasiparticles within the junction may be realized by making the contact leads from a superconductor with a higher value of Δ than the superconductor used for the junction. Since all quasiparticles created in the junction have energies close to the gap they have insufficient energy to get into the contact leads.

References:

1. D. McCammon, to appear in IEEE Trans. Nucl. Sci.

2. D. Hueber, C. Valette and G. Waysand: Cryogenics 21, 387 (1981)

3. A. Drukier, L. Stodolsky: Phys. Rev. D 30, 2295 (1984)

4. B. Cabrera, L.M. Krauss, F. Wilczek: Phys. Rev. Lett. 55, 25 (1985)

5. M.W. Goodman, E. Witten: Phys. Rev. D 31, 3059 (1985)

6. L. Gonzales-Mestres, D. Perret-Gallix: LaaPP-EXP-83-04

7. W. Seidel, L. Oberauer, F.v. Feilitzsch: accepted for publication in Rev. Sci. Instr.

8. E. Fiorini, T.O. Niinikoski: Nucl. Instr. and Methods 224 , 83 (1984)

9. N. Coron, G. Dambier, G.J. Focker, P.G. Hansen, G. Jegondez, B. Jonson, J. Leblanc, J.P. Moalic, H.L. Ravn, H.H. Stroke and O.Testard: Nature 314, 75 (1985)

10. T.O. Niinikoski and A. Rijllart: Europhys. Lett. 1 (10), 499 (1986)

11. C.A.Hamilton: Cryogenics 20, 235 (1980)

12. H.H. Andersen: Nucl. Instr. and Methods 12, 437 (1985)

13. V. Narayanamurti, R.C. Dynes, P. Hu, H. Smith, and W. F. Brinkman: Phys. Rev. B 18, 6041 (1978)

14. K.E. Gray: In Nonequilibrium Superconductivity, Phonons, and Kapitza Boundaries, ed. by K.E. Gray (Plenum Press, New York, 1981) p.131

15. M. Kurakado: Nucl. Instr. and Methods 196, 275 (1982)

16. A. Rothwarf, B.N. Taylor: Phys. Rev. Letters 19, 27 (1967)

17. W. Eisenmenger, K. Laßmann, H.J. Trumpp, and R. Krauß: Appl. Phys. 11, 307 (1976) and Appl. Phys. 12, 163 (1977)

18. P.W. Epperlein, K. Lassmann, and W. Eisenmenger: Z. Physik B 31, 377 (1978)

19. H. Kraus, Th. Peterreins, F. Pröbst, F.v. Feilitzsch, R.L. Mössbauer, V. Zacek, and E. Umlauf: Europhys. Lett. 1, 161 (1986)

20. D. Twerenbold and A. Zehnder: J. Appl. Phys. 61, 1 (1987)

21. D. Twerenbold: Phys. Rev. B 34, 7748 (1986)

The Possible Impact of Thermal Detectors in Nuclear and Subnuclear Physics

E. Fiorini

Dipartimento di Fisica dell'Universita' di Milano,
Istituto Nazionale di Fisica Nucleare, Sezione di Milano

1. Introduction

Historically the first use of a calorimeter in nuclear physics dates to 1903 when P. Curie and A. Laborde [1] used a calorimeter to reveal the energy produced by radioactivity. Another great contribution of calorimetry to fundamental physics was the experimental discovery that some energy was missing in beta decay of ^{220}Bi [2]. This fact was interpreted by W. Pauli in his famous letter of Dec. 4-1930 as due to the emission of a neutral particle later named neutrino by Enrico Fermi. In this brief and admittedly incomplete review I will summarize the present experimental status of a few problems in subnuclear physics where the "thermal approach" can lead to important and possibly unexpected results. I will limit myself to bolometers [3] which have been proved recently [4-8] to perform at low temperature as good and sometimes even fast [8] single-particle detectors. Many developments and plans on the "bolometric" and on the equally important "junction" approach have been presented at this conference.

Let me just remind one that in an "ideal" bolometer, acting as a pure insulator at low temperature, the heat capacity would be

$$C = 1944 \left(\frac{V}{V_m}\right)\left(\frac{T}{\theta_D}\right)^3$$

where V and V_m are the volume of the detector and the molecular volume, respectively and θ_D is the Debye temperature. The energy resolution, if all the energy of the incoming particle is deposited in form of heat, is expected to be

$$\Delta E = \xi \sqrt{K\ C \cdot T^2}$$

where k is the Boltzmann constant and ξ depends on the thermometer sensitivity and can range from one to ten. A value of three has been assumed for the sake of simplicity in the following considerations.

2. Double beta decay

Introduced by M. Goeppert Mayer since 1935 this peculiar process involves [9,10] a triplet of isobars (A, Z), (A, Z+1) and (A, Z+2) where single beta

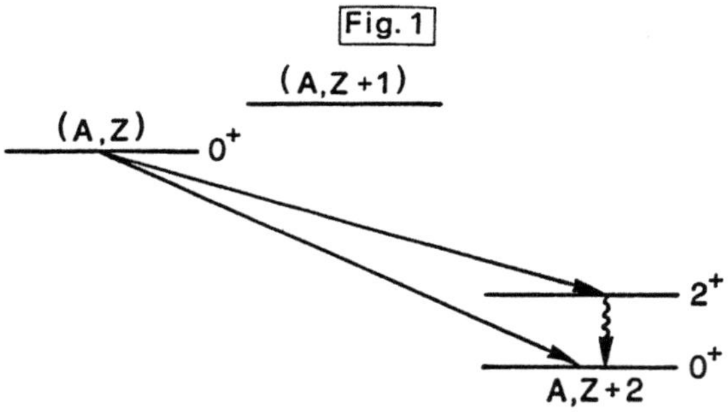

Fig. 1. Scheme of double beta decay to ground and first excited state of daughter nucleus.

decay to (A, Z+1) is forbidden or at least strongly hindered. (A, Z) could then decay (Fig. 1) into (A, Z+2) with the contemporary emission of two electrons in one of the following channels:

$$(A, Z) \rightarrow (A, Z + 2) + 2\,e + 2\bar{\nu}_e$$

$$(A, Z) \rightarrow (A, Z + 2) + 2\,e$$

$$(A, Z) \rightarrow (A, Z + 2) + 2\,e + X$$

In the first decay, which is allowed by all present laws of elementary particle physics, the transition energy is shared by four particles, the nuclear recoil energy being negligible. In the second decay (normally called neutrinoless) which is forbidden by the law of conservation of the lepton number, the two electrons share the total transition energy. Due to available phase space, the rate for this decay would be much larger than for the first one, thus allowing a very powerful means to search for violation of the lepton number. In the third process, also involving lepton number non conservation, a neutral Goldstone boson, named Majoron after the italian physicist Ettore Majorana, would be emitted and would share the transition energy with the two electrons.

The distribution of the sum of the two electron energies expected in the three processes is shown in fig. 2: it presents a sharp peak only in the neutrinoless mode.

Double beta decay has been studied with geochemical methods, searching in a rock of geological age, containing (A, Z), for an abnormal abundance of the isotope (A, Z + 2) cumulated in it by double beta decay of (A, Z + 2). These experiments were so far the only ones capable to give evidence for some type of double beta decay: ^{82}Se - ^{82}Kr and ^{130}Te - ^{130}Xe and possibly ^{128}Te ^{128}Xe. There are however some disagreements, and it is impossible with these methods to discriminate directly neutrinoless from two neutrino decays.

No evidence for any type of double beta decay has been obtained so far in the so called "direct" or "instrumental" experiments aiming at the detection of the

Fig. 2

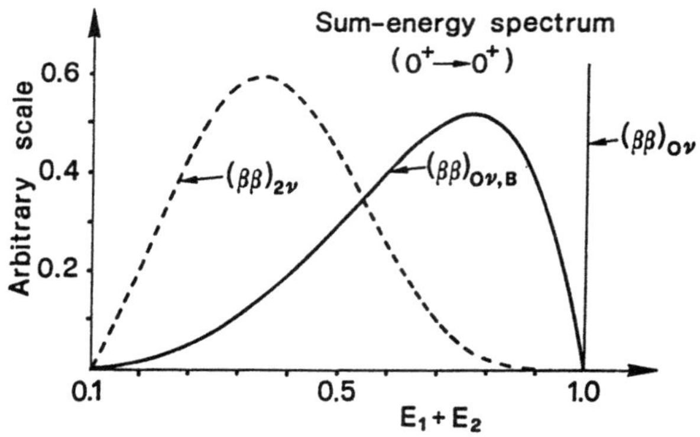

Fig. 2. Distribution of the sum on the two electron energies in the three double beta decay modes.

two electrons and possibly at the measurement of their energy. In some of these experiments the double beta decay material, usually in form of one or more thin sheets, is placed inside a particle detector in order to observe and possibly measure the electrons.

A different approach has been suggested since 1960 [11] and is more relevant to calorimetric detectors: it consists in the use of a detector which acts at the same time as source of double beta decay. The most stringent results [12] with direct methods have in fact been obtained with a technique initiated by the Milano group in 1967: the use as source and detector of a large Ge crystal, since natural Germanium contains 7.7% of ^{76}Ge which can double beta decay to ^{76}Se with a transition energy of 2,040.7 + .5 keV. Neutrinoless double beta decay would therefore appear as a sharp peak at this energy in the spectrum of the pulses from a well shielded and high-resolution germanium detector. A similar method presently being studied, with detectors already in operation in Italy and in USSR, consists in the use of ionization, proportional or time projection chambers filled with Xenon acting as source of double beta decay of 136 Xe. The negative results on neutrinoless double beta decay obtained with germanium detectors are impressive (a few 10^{23} years as lower limits for the half-lifetime) and a similar sensitivity is expected for the Xenon experiments, but further improvement seems hard due to the difficulty to construct very large detectors with good resolution. It is moreover essential to extend to other nuclei the search for double beta decay. This could be achieved [13, 14] with a large crystal of an ultrapure double beta decay active element in thermal contact (Fig. 3) with a small high-sensitivity thermometer, which could also be simply obtained by implanting in a small region of it. Even if the heat capacity of small implanted bolometers was found [4,6,8] to be larger by an

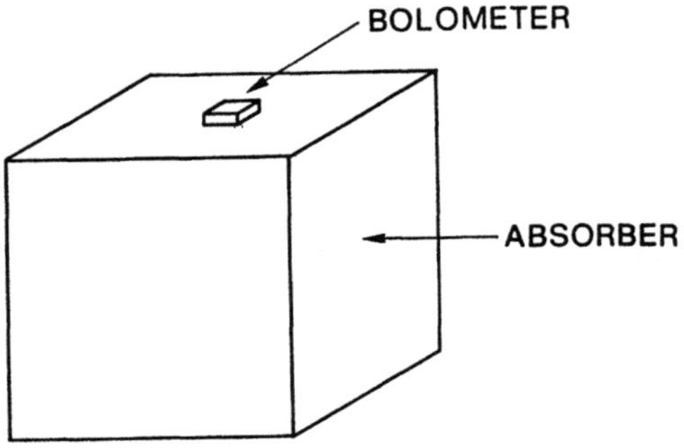

Fig. 3. A possible detector for double beta decay.

order of magnitude than expected, it would be anyway negligible with respect to that of the large double beta decay source. One should choose as possible candidates [13] elements with large nuclear transition energy and favourable nuclear matrix elements for double beta decay, and at the same time with low Debye temperature. Superconductors should in principle be avoided since part of the energy delivered by the two electrons could be lost in formation of Cooper pairs, but a low magnetic field could be used to suppress superconductivity. We report in Table I a few possible candidates for double beta decay where the energy resolution at T= 10 mK could be of the order of one keV, better than achievable with present solid state detectors at these energies. These values are obviously only indicative, and can be confirmed only with measurements on the specific heat of these materials at low temperature. This activity is presently going on in our laboratory in Milano.

Table I

Possible double beta decay detectors (operated at 20 mK with 1 keV resolution)

Isotope	Isotopic Aboundance (%)	Debye Temp (K°)	Transition Energy (keV)	Mass (Kg)
^{76}Ge	7.7	370	2040.7±.5	122
^{96}Zr	2.8	291	3356±7	74.5
^{100}Mo	9.6	460	3033±7	309
^{116}Ca	7.6	209	2808±6	34
^{124}Sn	6	195	2277±6	29
^{130}Te	34.5	153	2534±10	15
^{150}Ma	5.6	163	3366±8	21

3. Electron stability

Since no particle with charge lower than one has been found, the decay of the electron would probably imply non conservation of electron charge, which is automatically rejected by theorists, with some noticeable exception [15]. As an experimentalist I believe, however, that it should be tested anyway; one could just think, as an example, that the experimental evidence obtained by Cox in 1928 [16] on longitudinal electron polarization in beta decay was never taken seriously just because it violated parity!.

One could search for electron decay in the "exclusive" channel (γ + ν) where the photon energy should equal half of the electron mass. The corresponding line at 255 keV is broadened however by Doppler effect, since the electron decays in flight. This often prevents to take advantage of good resolution detector like Ge diodes. The best limit on electron decay in this channel has been obtained by F.T. Avignone et al. [17] and corresponds to 10^{25} years at the 90% confidence level.

A different, and model independent, approach is the search for the "inclusive" decay where the electron simply "disappears". If we assume for instance that such a decay occurs for a K-shell electron, the "hole" left by the electron would be filled by the outside ones giving rise to an electromagnetic cascade and therefore to the presence in the spectrum of a peak corresponding to the K-shell energy. The best limit on this type of decay, obviously less stringent than for the "esclusive" one, has been obtained by the Milano group with a Ge(Li) and corresponds to 2×10^{22} years [18].

Low-temperature calorimeters could provide an efficient tool to search for the inclusive decay of the electron, and also perhaps to improve the limit on the "exclusive" one. In this case the choice of the material is easier than for double beta decay: one could simply require a large Debye temperature and a reasonable energy for the K, or (if Z is sufficiently large) also for the L or M shell electrons. Some possible candidates are reported in Table II with the corresponding resolution at 10 m°K and the K-shell energy (or energies for compounds).

4. Other rare events

In addition to interactions of neutrinos or of dark matter particles [9,13,14] many other rare nuclear events could be studied with large calorimeters at low temperature, taking advantage of their good resolution. I would like just to mention rare nuclear decays, existence of natural isotopes with large lifetimes, detection of weak radioactive contaminations etc.. If the detector is very large, pile-up could become a problem.

Table II
Possible detectors for electron decay

Atom or molucule	K-shell energy (keV)	θ_D (°K)	C(at 20 mK) (J/K)	ΔE (eV)
C	285	2220	1.2×10^{-10}	15
Si	1.839	645	2.1×10^{-9}	63
Cr	5.99	630	1.2×10^{-9}	48
Mo	20.0	460	1.7×10^{-9}	57
Ru	22.12	499	1.2×10^{-9}	49
Rh	23.22	480	1.4×10^{-9}	51
W	69.52	405	1.3×10^{-9}	50
Re	71.68	430	1.1×10^{-9}	45
TiO_2	0.543 4.97	782	4.0×10^{-10}	28
GeO_2	.543 11.1	760	3.4×10^{-10}	26
TiB_2	.191 4.66	1010	2.2×10^{-10}	21

If we consider for instance a Germanium detector with a mass of a ton, the pulse risetime, even if simply due to acoustic propagation, can be as large as hundreds of microseconds and the decay time even larger. The counting rate at sea level without shield and in normal radioactive environment would be around half a million counts per second above 10 keV, and therefore absolutely inacceptable. Only underground and with a good shield against local radioactivity, can the background be reduced to the reasonable figure of a few tens of counts per second.

5. Neutrino mass

Even in searches for non zero electron neutrino mass, low-temperature calorimetry could provide considerable advantages, and in fact design studies have been presented by the Wisconsin [19], Genova [20] and Milano [21] groups. The present best sensitivities in measurement of antineutrino mass have been reached in the study of the Curie plot of tritium beta decay, where a mass in the range 17 to 40 eV has been suggested by the Moscow group. There is some, not

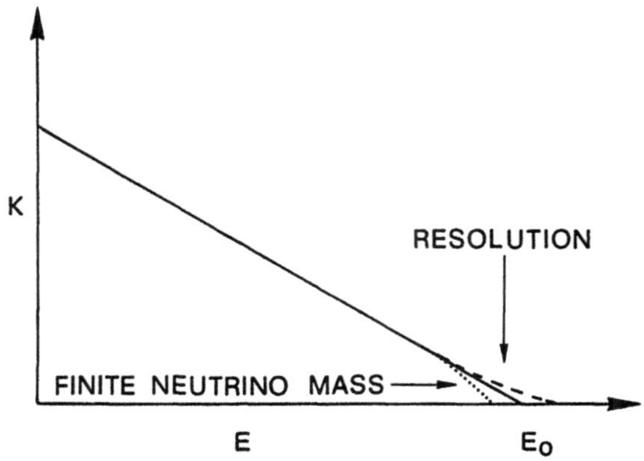

Fig. 4. Effects of the finite neutrino mass and of the resolution on the Curie plot. The low energy part of the spectrum is taken as unchanged due to the low population of the large energy region.

yet absolute, contradiction [22] with the negative results of the Zurich, Los Alamos and Tokyo (INS) groups which give upper limits of 18, 36 and 31 eV. Let me mention here the experimental difficulties of these searches:

a. the Curie spectrum expected for a zero neutrino mass and for an instrument with infinite resolution is a straight line which crosses the horizontal axis (Fig. 4) in a point corresponding to the transition energy. The presence of a massive neutrino would distort this spectrum in a direction opposite to the distortion due to the experimental resolution. An excellent knowledge of the energy resolution of the apparatus is therefore essential;
b. the end point energy is known with a precision much lower than the experimental resolution: it has therefore to be obtained from the fit to the spectrum as a third parameter, together with neutrino mass and experimental resolution;
c. the last part of the spectrum contains only a small fraction of the events: less than one per million and one per billion in the last 100 and 10 eV, respectively; it is therefore hard to reach a good statistic in the region of interest;
d. due to the weakness of the signal in the last region of the spectrum, the background could represent a considerable problem;
e. the decay of tritium can also occur to an excited state of ^{3}He, which further complicates the analysis;
f. the sensitivity of the experiment has to be around a ten of electronvolt: at this level molecular and atomic effects become important. Experiments should be carried out and compared with different tritium targets and possibly with tritium in gaseous, liquid and solid phase.

As an example of bolometric detector of neutrino mass [21] let us consider a diamond or silicon wafer of 1 x 1 mm^2 surface and .1 mm depth implanted on both sides with tritium at a depth around 5/μm. Theoretically the heat capacity at 10 mK could be as low as 10^{-17} to 10^{-14} J/K with an "ideal" resolution ranging from a few tens of meV to one eV. Recent experiments by McCammon et al. [4] indicate that at least the latter figure is not impossible to be reached. The real problem comes, at least in my opinion, from the pile-up of pulses which could contribute very dangerously to the background around the end point of the spectrum, due to its peculiar shape. If one assumes for instance a decay time of the order of 10_{μ}s [8] and a tritium activity of one Bq, the pile-up would be perfectly tolerable (3 x 10^{-3} events per year), but in a year of running time we would have only ~10 events in the last 100 eV of the spectrum and only ~0.01 in the last ten! One can think obviously of a mosaic of implanted wafers, each with its electronic chain, but gain equalization and other problems would make the experiment considerably complex.

6. Conclusions

I would like to conclude that bolometric detectors could provide excellent resolution, large masses of detecting materials, and good choice among various substances to investigate nuclear, atomic and molecular effects. This conference represents a step forward in transforming this "could" into "do".

I am indebted to my colleagues A. Alessandrello, D.V. Camin, A. Giuliani, C. Liguori and G. Pessina for many suggestions on various experimental points, and to L.B. Okun for theoretical discussions an electron stability.

References

1. P. Curie and A. Laborde: Compt. Rend. 136, 673 (1903).
2. C.D. Ellis and A. Wooster: Proc. Roy. Soc.(London) A117, 109 (1927).
3. I adopt here for "bolometer" the standard definition in elementary physics, namely: a thermometer, acting as microcalorimeter, such to measure very small changes of temperature due to absorption of energy.
4. D. McCammon et al.: I.E.E.E. in Nucl.Sci. 33, 236 (1986), also for previous references.
5. N. Coron et al.: Nature 314, 75 (1985).
6. J.C. Overley et al.: Nucl. Instr. and Meth. B 10/11, 928 (1985).
7. C. Boragno et al.: Cryogenics, 24, 681 (1984).
8. A. Alessandrello et al.: Heat capacity of low-temperature Ge- and Si-calorimeters and optimization of As-implanted silicon thermistors, Nucl. Instr. and Meth. (in press); T.O. Niinikoski et al. (CERN-Milano Coll.) Europhys. Lett. 1, 499 (1986).
9. E. Fiorini: Experimental perspectives in low-energy lepton physics, in Nuclear Beta Decays and Neutrino, ed. by T. Kotani, H. Ejiri and E. Takasugi, p.11 (1986) (Introductory review).

10. W.C. Haxton and G.J. Stephenson Jr: Progr. in Part. and Nucl. Phys. 12, 409 (1984), and M. Doi, T. Kotani and E. Takasugi: Prog. of Theor. Phys. Suppl. No. 83, 1985.
11. G.F. Dell'Antonio and E. Fiorini: Suppl. Nuovo Cim. 17, 132 (1960).
12. D. Caldwell: Review of double beta decay results with 76 Ge - Neutrino '86, ed.by. T. Kitagaki and H. Huta, p. 77, 1986.
13. E. Fiorini and T.O. Niinikoski: Nucl. Instr. and Meth. 224, 83 (1984) also for previous references.
14. B. Cabrera, D. Caldwell, and B. Sadoulet: Low- temperature detectors for neutrino experiments and dark matter searches, preprint.
15. L.B. Okun and Y.B. Zeldovich: ITEP Report No. 79, 1978; L.B. Okun: Lepton and Quarks, ed. by North Holland, Amsterdam (1982), p. 181.
16. E. Segre': Nuclei and Particles, ed. A. Benjamin, New York.
17. F.T. Avignone et al.: Phys. Rev. D34, 97 (1986) also for previous references.
18. E. Bellotti et al.: Phys. Lett. 124B, 435 (1983).
19. D. McCammon et al.: A New Technique for Neutrino Mass Measurements, preprint.
20. A. Blasi et al.: Una Nuova Determinazione di Limiti di Massa per l'Antineutrino, preprint.
21. Milano Group: Proposal for the Thermal Determination of the Neutrino Mass, preprint.
22. T. Ohshima: Review of Neutrino Mass Measurement, INS-REp.-598, July 1986, preprint.

Considerations on Front End Electronics for Bolometric Detectors with Resistive Readout

A. Alessandrello, **D.V. Camin**, A. Giuliani, and G. Pessina

Istituto Nazionale di Fisica Nucleare, Via Celoria 16, I-20133 Milano, Italy

1. Introduction

We have started some time ago a research activity to evaluate the possible use of cryogenic detectors in the detection of very rare events [1]. First tests of silicon calorimeters have been performed at CERN and showed the feasibility of using such devices as photon detectors. Germanium, totally doped commercial thermometers have been also used as detectors for alpha particles [2], [3]. In collaboration with SGS, the main local semiconductor manufacturer, we are currently developing a series of silicon cryogenic detectors which will be tested with low-energy photons emitted by a 55Fe X-ray source.

Only at the beginning of February 1987 a dilution refrigerator from Oxford Instruments was installed in Milan [4]. This machine which includes a top-loading option, reached a lowest temperature of 4.7mK.

After commissioning of the refrigerator, our group started an intense activity of detector characterization. Measurements of thermistor resistance vs. temperature, and voltage-current load lines have been performed.

Also an active research on the front end electronics for this type of detectors started at the beginning of 1986.

This report will concentrate mainly on the problem of signal amplification in presence of noise from the detector and active devices.

2. Bolometers characteristics

A bolometer consists of an absorber and a thermometer in contact with it.

The absorber is made out of a pure crystal, typically silicon or germanium.

One part of it is doped with a proper dopant concentration to form the thermistor. Also the whole crystal may be doped so that the thermistor and the absorber share the available volume.

When a particle deposits all or part of its energy into the absorber after a process of phonon production and further thermalization, the heat created will raise the bolometer temperature. If the heat capacitance is low, condition which is met at cryogenic temperatures according to the Debye law, the temperature rise will be measurable through the corresponding resistance change in the thermistor.

A bias circuit will provide a current so that a change of voltage will be measured.

A simple calculation shows that the amplitude of the voltage is given by:

$$V_O = I_B \cdot R_T \cdot \frac{A}{TC} \cdot E \qquad (1)$$

where I_B is the bias current, R_T is the parallel combination of the biasing resistor and the thermistor resistance, A is the thermistor sensitivity expressed as d(logR)/d(logT), C is the heat capacitance and E is the deposited energy.

Figre 1 shows a typical bolometer and the biasing circuit.

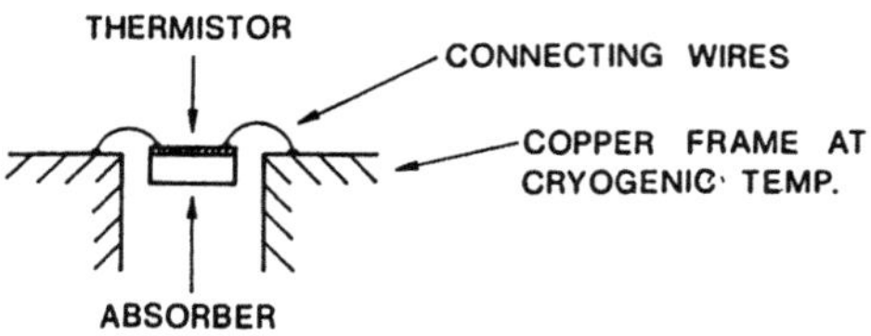

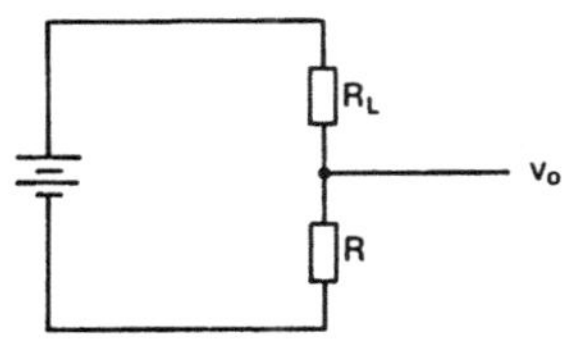

Fig. 1: Typical bolometer and biasing circuit.

The wires connecting the thermistor to the biasing circuit will also conduct the heat absorbed so that after some time the thermal equilibrium is again reached.From the thermal model, depicted in Fig. 2, it is possible to calculate the time structure of the voltage pulse. If the thermalization time is small compared to the physical time constant τ (τ =C/G where G is the thermal conductance

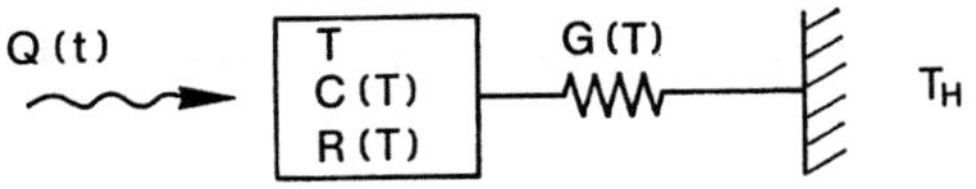

Fig. 2: Bolometer thermal model.

of the connecting wires), then the pulse will show exponential decay and amplitude given by (1).

The effective time constant τ_e will be different from τ and will depend on the operating point as will be explained later. The static characteristic voltage vs. current of a bolometer with negative thermal coefficient is shown in Fig. 3.

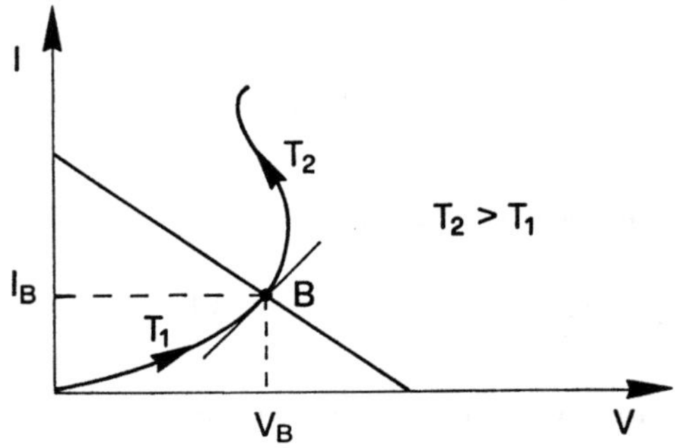

Fig. 3: Current-voltage characteristics of a negative bolometer

Due to self-heating the V-I load line is not a straight line, but shows rounded shape. There are a family of curves, one for each heatsink temperature. The tangent at the origin corresponds to the resistance at the "cold" condition, that is to say when self-heating is negligible. Main parameters obtained from the load line are R, static resistance, Z dynamic impedance which also depends on the frequency, and the bias current I_B for every point.

Determination of these parameters allows the calculation of the thermal conductance as:

$$G = \frac{(Z + R)}{(Z - R)} \frac{A\, I_B^2\, R}{T}$$

We have developed in Milano a system based on a personal computer which automatically determines the load line of our bolometers and calculates the main parameters. Fig. 4 shows one of the curves obtained with our system.

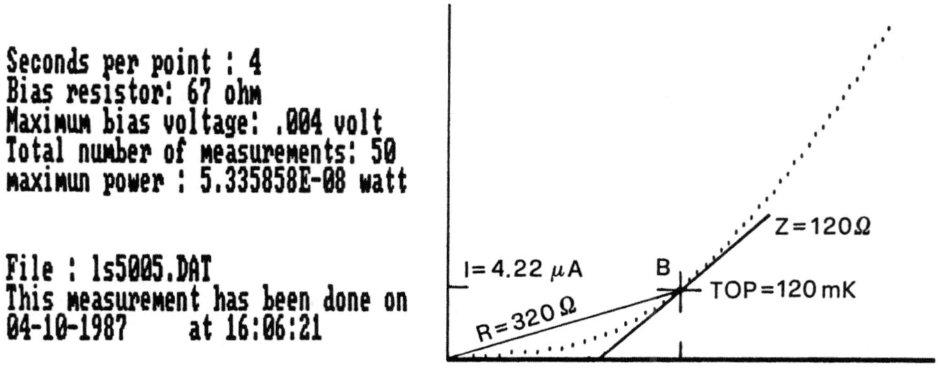

Fig. 4: Load line of a Ge bolometer.

The responsivity of the bolometer $S(\omega)$ can be expressed in terms of the above-mentioned parameters as

$$S(\omega) = \frac{V_o(\omega)}{Q(\omega)} = \frac{1}{2I_B} \cdot \frac{Z/_R - 1}{Z/_{R_L} + 1} \cdot \frac{1}{1 + j\omega\tau e}$$

where

$$\tau_e = \tau \cdot \frac{Z + R}{2R} \cdot \frac{R + R_L}{Z + R_L}$$

is the effective time constant of the system.

If the thermalization time is negligible, then:

$$Q(t) = E \cdot \delta(t) \qquad \text{and} \qquad v_o(\omega) = S(\omega) \cdot Q(\omega) = S(\omega) \cdot E$$

In this case an exponential pulse of amplitude $E\ S(0)/\tau_e$ and decay time constant τ_e will be developed.

The expression of the pulse amplitude in terms of the electrical parameters:

$$v_o(o) = \frac{S(o) \cdot E}{\tau_e} = \frac{1}{2I_B} \cdot \frac{Z/_R - 1}{Z/_{R_L} + 1} \cdot \frac{E}{\tau_e} \qquad (2)$$

is totally equivalent to (1) expressed before in terms of the physical parameters A,C and T.

From (1) it is also possible to observe that a pulse will be developed only when the biasing point is in the rounded part of the curve where $Z \neq R$.

Returning to Fig. 4 we can see that biasing in point B we get R=320 ohm, Z=120 ohm, G=2 exp-7 W/K, C(lattice)=1 exp-10 J/K,

τ=C/G= 0.5 ms, S(0)=-6.6 exp-5 W/K, then for 6MeV alpha particle we get v_o=0.16 mV, τ_e=0.4ms.

The resolution of this type of detectors may be very high. In fact the theoretical limit is expressed by (3):

$$\Delta U_{rms} = (K_B T^2 C)^{1/2} \quad (3)$$

where ΔU is the r.m.s. value of the spontaneus energy fluctuation of the detector of heat capacitance C held at temperature T, K_B is the Boltzmann's constant. The actual theoretical limit will be higher and depends on the operating point [5]. In this case (3) has to be multiplied by a factor ξ typically ranging between 1 and 2.

Unfortunately not all the energy is converted into heat, but part of it produces ionization, or defects in the crystal or simply escapes from the detector [5]. The final resolution is lower than the theoretical limit. To our knowledge, the best results obtained so far with this type of detectors were achieved by Mc Cammon's group [6], [7], who reached 35 eV for the 5.6 KeV line of a 55Fe X-ray source.

3. Signal analysis

The classical theory developed by Jones [8] shows that the equivalent circuit of a bolometer made with a semiconductor material (which exhibits negative temperature coefficient) is the one depicted in Fig. 5.

There, equations (4),(5) and (6) are valid where e_g is the voltage generator

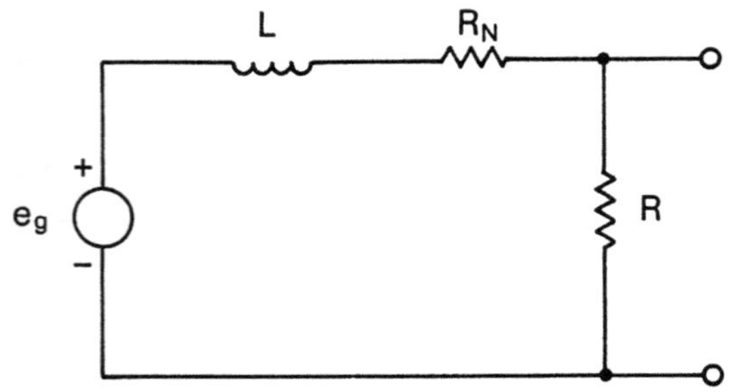

Fig. 5: Equivalent circuit of a negative bolometer.

$$e_g = -\frac{1}{2I_B}Q \qquad (4)$$

$$R_N = -\frac{R}{2}\left(\frac{K}{\alpha P} + 1\right) \qquad (5)$$

$$L = -\frac{R}{2}\frac{K\tau}{\alpha P} \qquad (6)$$

representing the signal source, R is the static bolometer resistance, I_B the bias current, P the bias power, K the thermal conductance, α the thermal coefficient, and τ the thermal time constant.

It is then evident that whenever R is relatively high the capacitance of the cabling carrying the signal to the preamplifier will impose a high - frequency cut-off which will deteriorate the pulse amplitude, creating a ballistic deficit dependant on the pulse risetime. Putting some numbers, if R=10MΩ and C_i = 150 pF the high frequency cut-off will be only 100 HZ. This situation has to be avoided and so we decided to take the following actions: to use a low noise room temperature electronics with low resistance bolometers and putting very close to the detector the first preamplifier stage when the bolometer resistance is high. The last action imposed us the selection and the study of the performance of a suitable solid state device which might work close to the detector, therefore at very low temperature.

4. Operation of solid state devices at low temperature

4.1. Conduction mechanism

In selecting suitable solid state devices for low temperature we have to consider first the three commonly available materials: silicon, germanium, and gallium-arsenide. Figre 6 shows a band diagram of a semiconductor material.

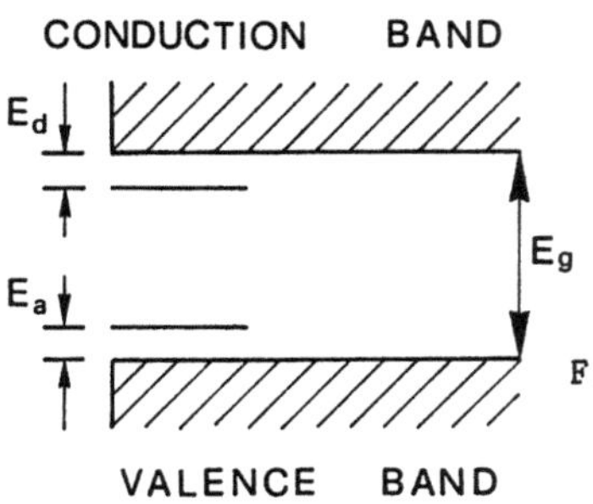

Fig. 6: Energy band diagram of a semiconductor material.

Table I shows the value of E_g, E_d and E_a energy gap, donor and acceptor level respectively for Ge, Si and GaAs.

Tab. I: Value of the energy gaps for Ge, Si, GaAs.

	Si	Ge	Ga-As	
Eg(330°K)	1.12	0.66	1.42	eV
Eg(0°K)	1.17	0.73	1.52	eV
donor:	P, As, Sb		S, Sn, Se	
Ed	40-54	10-13	6	meV
acceptor:	B, Al, Ga		Be, Mg, Zn	
Ea	45-72	10	30	meV

In n-type intrinsic semiconductors the electron density is given by:

$$n \div \exp(- E_d/2KT) \qquad (7)$$

This equation shows that at very low temperature there will be a carrier frozen-out which depends exponentially on E_d .Looking at the values of E_d in table I we see that Ge and GaAs will show frozen-out of carriers at a much lower temperature than silicon. Actually Ge FETs have been used years ago for low - temperature operation [9], [10].

Another parameter to be taken into account when considering the operation of solid state devices at low temperature is the behaviour of the conductivity σ which, for a n-type semiconductor is expressed by:

$$\sigma \simeq q \mu_n n \qquad (8)$$

It becomes evident that besides the carrier concentration, the mobility dependance with temperature affects the operation at low temperature. For instance, conduction in GaAs at 4.2 K is mainly limited by mobility [11].

The very well known fact that there is an optimum operating point for a Si JFET at about 120 K is due to the combined effect of mobility increasing and carrier density decrease with lowering the temperature [12].

The use of minority carrier devices like bipolar transistor is not advisable as the reduction of carrier life time [11] creates a reduction of the diffusion lenght and a consequent decrease in the collector current and an increase of the base current, with a final result of getting a much lower h_{fe} current amplification factor.
MOS FETs operate well at LHe temperature because frozen-out of carriers is compensated by the high electric field due to the thin oxide layer existing between gate and channel.

4.2. Noise

Noise sources in field effect transistor are mainly: thermal noise in the channel and generation-recombination noise. The spectral power density of the first source is given by:

$$\bar{e}^2_{nTH} = 4KT \frac{\gamma}{g_m} \qquad (9)$$

Going from room temperature to an optimum temperature (typically 120 K) there is a noise reduction. Below the optimum temperature g_m will decrease, γ increases heavely and the net effect is a high noise increase.

The low-frequency noise is produced by a mechanism which involves trapping and release of carriers due to the existance of impurities (i.e. gold in silicon).This process is characterized by a time constant τ [12].The spectral power density is given by [13]:

$$\bar{e}^2_{gr} \div \frac{\tau}{1 + \omega^2 \tau} \qquad (10)$$

where τ is the trap time constant. The increase of τ at low temperature is verified experimentally [14].

MOS FETs show a high value of A_f , the 1/f noise coefficient due to the presence of a high number of traps at the oxide semiconductor interface, which make these devices in principle too noisy for low frequency applications [15]. There is however a series of dual-gate

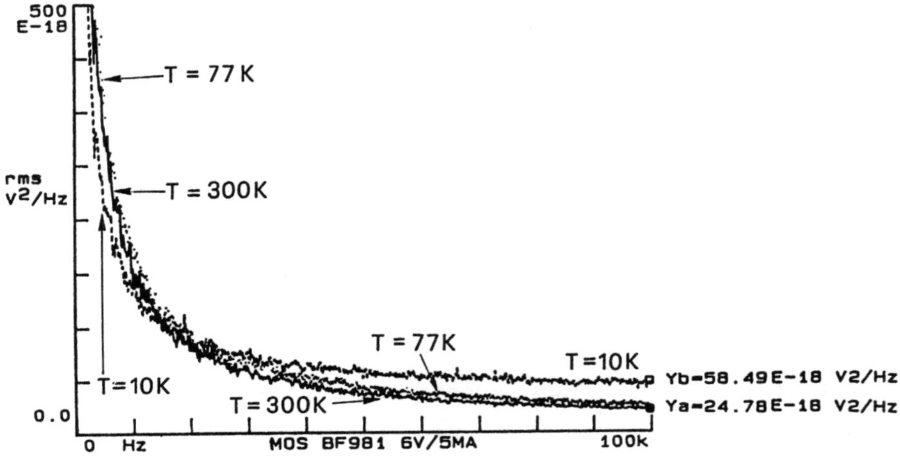

Fig. 7: Spectral power density of a BF981 MOS FET.

MOS FETs from Philips which exhibit an A_f value (1 exp-12 V^2) one or two order of magnitude lower than an average MOS FET even at low temperature. We have measured such devices. The results are given in Fig. 7.

Gallium Arsenide MES FETs have shown a favourable dependance of low-frequency noise with temperature. To our knowledge this is the only device which shows a decrease of A_f with lowering temperature.

We have measured down to 10 K and the results look promising, see Fig. 8.

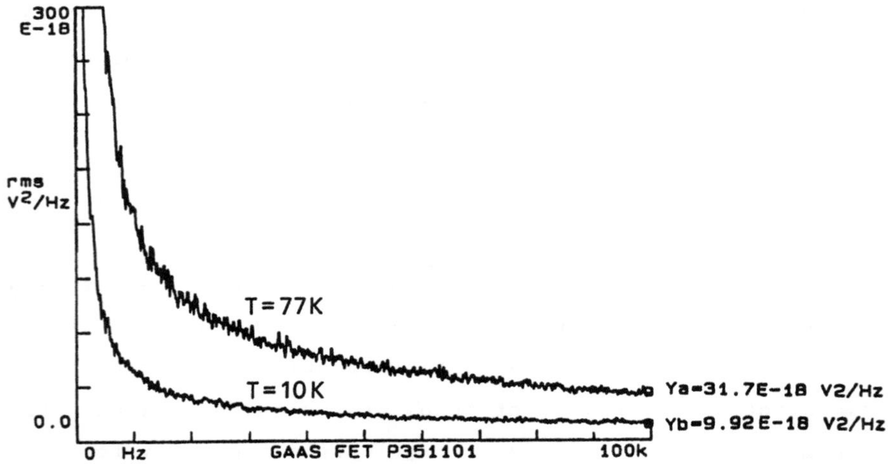

Fig. 8: Spectral power at different temperatures of a GaAs MESFET.

The mechanism controlling this process is still unknown. We have obtained the best results with a long gate device, the P35-1101 from Plessey. We will use this device in the front end electronics of the high-resistance bolometers.

5. Signal amplification

Fig. 9 shows the classical circuit configuration for signal readout.

The total noise at the input of the amplifier will be the sum of the contributions due to its input device and the electric equivalent noise representing the energy fluctuation of the bolometer itself. The latter can be represented by a Johnson's noise source kept at T_n, the so called noise temperature, which depends on the operating point (T_n is about 2.Tsink) [16]:

$$\bar{e}_B^2 = 4KT_n R_B \tag{11}$$

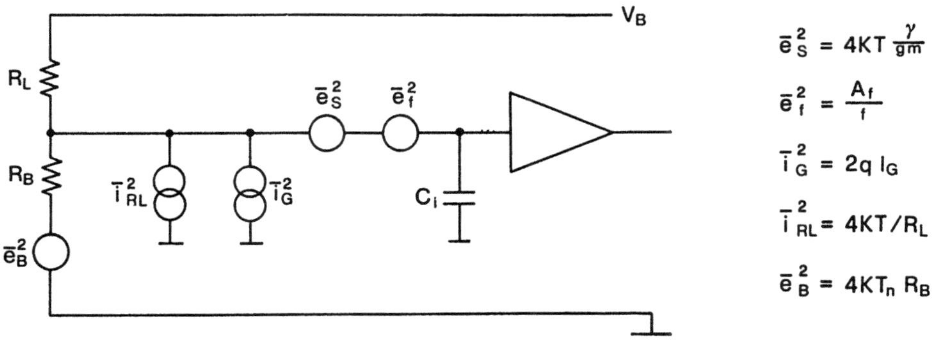

Fig. 9: Signal readout circuit with noise sources put into evidence.

It was mentioned above that the first stage of the preamplifier may be put near the detector to avoid the cable parasitic capacitance. This is particulary necessary when the bolometer resistance is high. A suitable device capable of operating at cryogenic temperature and exhibiting low noise has to be selected.

When the bolometer resistance is low then the effort has to be put in lowering the amplifier's noise contribution (see (11)). As the parasitic capacitance does not represent in this case the major problem, the front end electronics can work at room temperature, which permits to obtain a better noise performance.

We have developed an ultra-low noise differential preamplifier which is currently being used in the signal amplification of a Ge bolometer.

The main features of this preamplifier are: 400 pV/ √HZ differential input noise, 67 dB differential gain and 100 dB CMRR [17]. Figure 10 shows the spectral power density in the low frequency region whereas Fig. 11 shows the white noise region of this preamplifier.

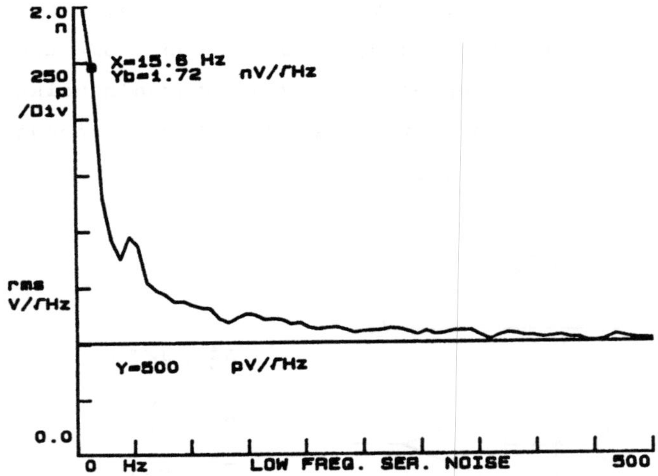

Fig. 10: Low-frequency noise of the voltage-sensitive differential input preamplifier.

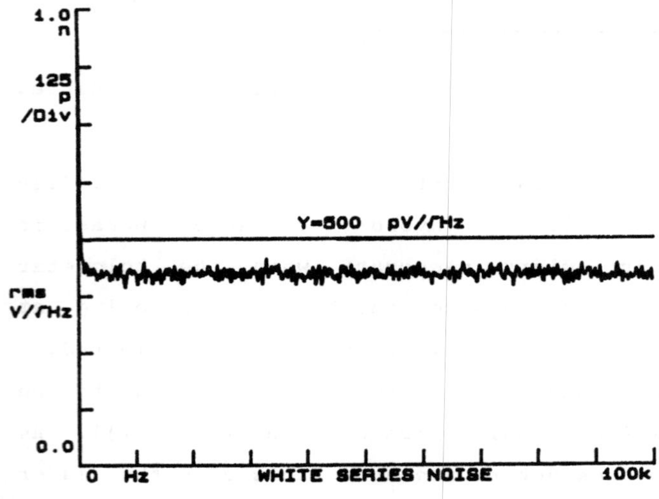

Fig. 11: White series noise of the voltage-sensitive differential input preamplifier

6. Conclusions

The continuous improvement of the bolometric detectors with resistive readout stresses the need of developing suitable low-noise front-end electronics.

This requires the research on new components capable of operating at cryogenic temperatures.

In this respect our results show that long gate GaAs MESFETs behave quite well at temperature near 1 K. Our future steps foresee a complete amplifier made using GaAs MESFETs working at 1 K.

Still, low-resistance bolometers permit the use of a much more comfortable room-temperature electronics.

7. Bibliography

[1] E. Fiorini. T.O. Niinikosky, N.I.M 224, 83 (1984)

[2] T.O. Niinikosky, A. Rijllart, A. Alessandrello, E. Fiorini, A. Giuliani.Europhysics letters, 1 (10), 499-504 (1986).

[3] A. Alessandrello, D.V. Camin, G.F. Cerofolini, E. Fiorini, A. Giuliani, C. Liguori, L. Meda, T.O. NIniikosky, A. Rijllart. Heat capacity of low-temperature Ge and Si calorimeters and optimization of As-Implanted silicon thermistor. To be published by Nuclear Instruments and Methods

[4] Oxford Instruments ltd.,Model 200 TLE.

[5] S.H. Moseley, R.L. Kelley, D. Mc Cammon, Journal of Applied Physics, 56, 1257 (1984).

[6] D. Mc Cammon, S.H. Moseley, J.C. Mather, R.F. Mushotzky, J. Applied Physics, 58, 1263 (1984).

[7] S.H. Moseley, R.L. Kelley, J.C. Mather, R.F. Mushotzky,A.E. Szymskowiak, D. Mc Cammon. Thermal detectors as single photon x-ray spectrometers. To be published by IEEE Transc. Nucl. Science.

[8] R.C. Jones, J. Optical Society of America, 43 (1), 1 (1953).

[9] E. Elad, M. Nakamura, IEEE Tran. Nuclear Science, 15, 283 (1968)

[10] E. Elad, M. Nakamura, IEEE Tran. Nuclear Science, 15, 477 (1968)

[11] B. Lengeler, Cryogenics, 14, 439, 1974

[12] W. Shokley, Proc. of IRE, 46, 973 (1958)

[13] C.T. Sah, Proc. of IRE, 52, 795 (1964)

[14] J.W. Haslett, E.J.M. Kendall, IEEE Tran. on Electron Devices, 19, 943 (1972)

[15] A. Van Der Ziel, Advance in Electronics and Electron Physics, 49, 225 (1979)

[16] J.C. Mather, Applied Optics, 23, 584 (1984)

[17] D.V. Camin, G. Pessina, P.F. Manfredi
Voltage-sensitive differential input preamplifier with outstanding noise performance. To be published by Alta Frequenza.

Coherent Neutrino-Nucleus Elastic Scattering in Ultralow-Temperature Calorimetric Detectors

T.O. Niinikoski and A. Rijllart
CERN, CH-1211 Geneva 23, Switzerland

We speculate on the measurement of the coherent forward peak in the neutrino-nucleus elastic scattering, using ultralow-temperature calorimetric techniques for the determination of the recoil energies down to below 1 keV. The detector would consist of an array of relatively large single crystal calorimeters made of various elements and compounds, cooled to below 100 mK temperature. The detector should be surrounded by veto track detectors for the elimination of the background events due to cosmic rays and neutrons and charged beam particles. The coherent event rate for some existing or projected neutrino beams and sources is calculated for a Ge detector.

1. Introduction

The neutral-current elastic scattering of neutrinos on nuclei is predicted to have a pronounced forward peak in analogy to electron scattering. For momentum transfer small compared with inverse target size, the relative phase factors of the waves scattered from the individual constituents of the target are small, and the waves add coherently [1] to make the amplitude

$$F(k',k) \sim G\, e^{-q^2R^2} \times$$
$$\times [Z(a_0+a_1/2)+N(a_0-a_1/2)+(Z_\uparrow-Z_\downarrow)(b_0+g_V b_1/2)+(N_\uparrow-N_\downarrow)(b_0-g_a b_1/2)]\times$$
$$\times \overline{u}(\nu')\gamma^0 u(\nu) \qquad (1)$$

Here a_0, a_1, b_0 and b_1 are the general phenomenological parameters of the electroweak interactions, which take the values $a_0 = -\sin^2\theta_W$, $a_1 = 1 - 2\sin^2\theta_W$, $b_0 = 0$ and $b_1 = -1$ in the Weinberg-Salam (WS) model. The spin dependent terms in (1) are small for nuclei with large mass and low spin and are exactly zero for nuclei with zero spin. There are several other models which cannot be ruled out as yet, although the WS model is the most popular one.

The range of momentum transfers where the coherence occurs is determined by the nuclear size R and is around $|q| < 100$ MeV/c; for heavier nuclei this corresponds to such small recoil energies that the events are below the threshold in any existing detector of reasonable size.

A new detector, based on thermal calorimetry in pure dielectric diamagnetic single-crystal materials, shows promise for reaching good energy resolution and threshold even in rather massive combined target-detectors.

The principle of the thermal detection of particles relies on the conversion of the deposited energy into heating of the phonons in the material, and has been proposed already some time ago [2]. The first evidence for the detection of cosmic rays in resistance thermometers was in 1975 [3], but more serious development work concentrating on single crystal Si and Ge detectors has been recent [4, 5,6,7]. McCammon et al. [8] have reached a spectacular 35 eV resolution in their Si detector using 6 keV x-rays.

2. Neutral current scattering

2.1 Differential cross section

The coherent scattering of neutrinos on heavy nuclei has several appealing features which may enable the study of the gauge field theories underlying the electroweak interactions. Let us consider a nucleus with Z protons and N neutrons, and apply the WS model in Eq. (1) ignoring the small spin-dependent terms. The differential cross section [9] in the forward elastic peak due to neutral current is then

$$\frac{d\sigma}{dq^2} = \frac{G^2}{8\pi}[Z(1 - 4\sin^2\theta_W) - N]^2 e^{-2bq^2} [1 - q^2 \frac{2ME_\nu + M^2}{4M^2E_\nu^2}] \quad (2)$$

and is independent of the neutrino type. Here θ_W is the Weinberg angle, G the Fermi constant, M the nuclear mass, E_ν the neutrino energy , and $b \cong r^2/6$ is related to the rms nuclear radius r in analogy with electron scattering [9]. The cross section has an enhancement roughly proportional to N^2 compared with charged current cross sections. This is strictly true for spin zero nuclei; other nuclei have additional small contributions from axial vector currents. These contributions depend more on the nuclear form factors but they are well below 10% of Eq. (2) [10] for medium heavy nuclei such as Fe.

The coherent enhancement of the cross section (2) occurs in the kinematic range $q^2 < (2b)^{-1} = 3/r^2 \approx (100\ \mathrm{MeV/c})^2$. At larger momentum transfers the situation is complicated by the quasielastic scattering, where the nucleus is left in an excited or broken state. There are, however, cases where simple analysis might be possible; some of these will be briefly discussed below. In the kinematic region where the coherent scattering dominates, we have $M^2 \gg q^2$ and the equality $q^2 = 2MT$ holds, where T is the laboratory kinetic energy of the recoil nucleus. For the purpose of rate estimates, we approximate $\sin^2\theta_W = 0.25$ and rewrite Eq. (2)

$$\frac{d\sigma}{dT} \approx \frac{G^2MN^2}{4\pi}[1 - \frac{T}{T_{max}}] e^{-T/T_{coh}} \quad (3)$$

where $T_{max} = 2E_\nu^2/(M + 2E_\nu)$. We wish to emphasize that Eq. (3) is supposed to be valid only up to $T_{coh} \approx 3/(2Mr^2)$ ($\approx$ 240 keV for Si; see table 1) whereas for $E_\nu \approx 40$ GeV in SPS neutrino beam and for M(Si) = 26 GeV, we have the hypothetical $T_{max} \approx 30$ GeV.

Fig. 1 shows the qualitative behaviour of the recoil spectrum for a Si target-detector. The upper line is the coherent cross section for a pointlike nucleus (at $E_\nu > 1$ GeV); the exponentially dropping line shows the cross section of Eq. (2) using the rms nuclear radius from electron scattering data [12]. The shaded area indicates the range of energy deposits in the detector where contributions from excited nucleons could add features both in the recoil spectrum and in the signal shapes.

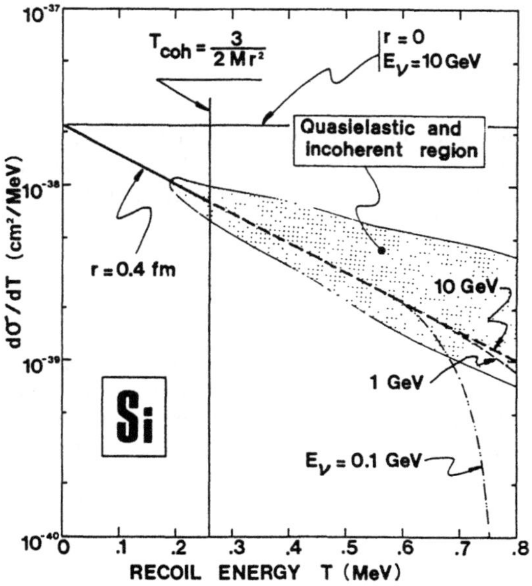

Figure 1: Differential cross section for coherent elastic neutrino-nucleus scattering

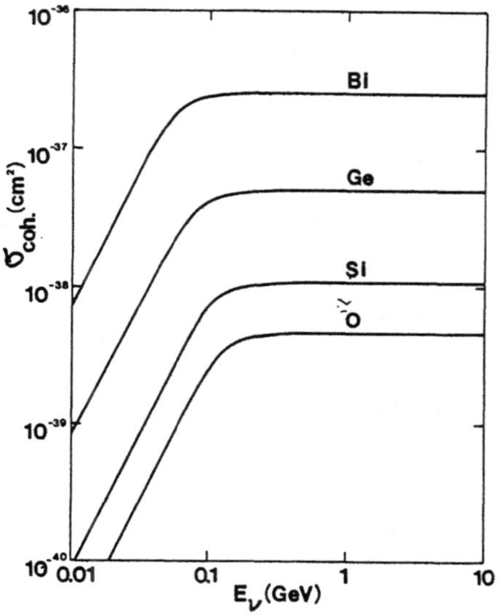

Figure 2: Total cross section for coherent neutrino scattering

2.2 Total coherent cross section

In estimating the total coherent cross section σ_{coh} we integrate the Eq. (3) to get

$$\sigma_{coh} = \frac{G^2}{4\pi} MN^2 T_{coh}\, f(\xi), \tag{4}$$

where $f(\xi) = 1 - (1 - e^{-\xi})/\xi$, and $\xi = T_{max}/T_{coh} = (2/3)E_\nu{}^2 r^2/(1 + 2E_\nu/M)$. In the high-energy limit we have $f(\xi) = 1$ and the cross section is constant and a few orders of magnitude higher than the usual neutrino cross sections. In the low energy limit $T_{coh} > T_{max} = 2E_\nu{}^2/M$, which implies $E_\nu < 50$ MeV, and the total coherent cross section decreases strongly with energy [13]:

$$\sigma_{coh} = \frac{G^2}{4\pi} N^2 E_\nu{}^2 \tag{5}$$

The Fig. 2 shows the behaviour of the coherent cross section for O, Si, Ge and Bi targets as a function of neutrino energy E_ν.

2.3 Quasielastic scattering

Above $q^2 \approx 3/r^2$ the recoil energy spectrum registered by the calorimeter would reflect the sum of many possible processes, one of which is the elastic coherent process

$$\nu + A \rightarrow \nu' + A$$

which goes down approximately as $\exp(-2b/q^2)$, and others are

$\nu + A \rightarrow \nu' + A^*; \qquad A^* \rightarrow A + \gamma$

$\nu + A \rightarrow \nu' + B + X; \qquad .....$

where the excitation energy of the quasielastic reaction products may show up delayed depending upon the lifetime of the particular state selected. The delayed calorimetric signal might provide a sensitive trigger for the selective study of some of these reactions and allow good discrimination among the various gauge models [11].

3. The detector

The development work at CERN has concentrated on the understanding of the thermal behaviour of the calorimeter, in order to be able to optimize the construction and to be able to extrapolate the detector size to the region of interest for neutrino experiments and other applications where large masses might be desirable. Our results on small integrated phosphorus-implanted Si detectors and melt doped Ge detectors are encouraging; these are discussed elsewhere in these Proceedings [14] in the light of the recent analysis of our previous data.

3.1 Calorimeter elements

For the purpose of the experiments discussed here, we need the total detector masses in the range 1 to 10^3 kg, consisting of .1 to 1 kg lumps of pure Si and/or Ge. The single crystal rods would be cut into slices and mounted into a frame cooled by a $^3He/^4He$ dilution refrigerator. The thermistors are attached to each detector by laser welding or sintering techniques, and are cut, etched and wired after attachment. Crystals of B and Bi, and also compounds such as BeO, Al_2O_3, TiO_2, GeO_2, SiC, and TiB_2 could be included in the detector, depending on the availability of pure single crystal materials and on the scope of the experiments.

Table 1: Calorimeter Material Properties.
The ΔE is given for 0.1 kg mass at 30 mK temperature.

Element	Z	M (g/mol)	T_{coh} (keV)	θ_D (K)	ΔE (eV)	σ_{coh} (10^{-38} cm^2)
C	6	12.000	892	2240	11	0.151
Si	14	28.086	242	647	46	0.527
Ge	32	72.590	54	378	64	2.526
Bi	83	208.980	10	119	212	12.777

The Table 1 gives the energy resolution of .1 kg calorimeter units at 30 mK temperature, basing on the uncertainty principle of quantum statistics

$$\Delta E_{rms} = \zeta\sqrt{(kT_0^2C_0)} \qquad (6)$$

where T_0 is the detector temperature, C_0 the heat capacity, and $\zeta \cong 2.5$ depends slightly on thermistor parameters; the table is based on the Debye heat capacity only.

3.2 Amplifiers

The preamplifiers must have their input stages near the detectors, in order to reduce the parasitic lead capacitance, and to protect against interference, cross coupling and noise. This allows to obtain a frequency response matching the counting rate, which is dominated by cosmic-ray muons (about 1.5 s^{-1} in 1 kg units).

The input stages of the preamplifiers can use either Si or GaAs JFET's cooled to 100 K or 10 K, respectively. The subsequent stages of amplification may use commercial integrated circuits; the amplifiers can be located at room temperature.

3.3 EMI control

The extremely small signals corresponding to temperature increments below 1 μK in the calorimeters require an ultimate in the control of the electromagnetic interference (EMI) in the difficult range below 1 MHz. The main sources of interferences are related to the operation of heavy electrical equipment and also to acoustic microphony. Our experience shows that by enclosing the equipment in a double-walled Faraday cage and heavily filtering the main power feed, these interferences can be reduced to an acceptable level. This technique is also efficient in removing interferences in the higher frequency spectrum, which cause direct heating of the thermistors and thermometers in the level 10^{-10} – 10^{-12} W; this heating fluctuates and adds to the noise of each detector in a correlated way. The severity of this heating may be visualized by recalling that the heat conductance is around 10^{-11} W/K from a calorimeter at 30 mK to the heat sink.

The data acquisition electronics must provide signal triggering, vetoing, digitization and transmission via fibreoptic links to a computer located outside the Faraday cage. Apart from the power feed into the cage, no other galvanic transmission line is allowed to penetrate the shield. It is also necessary to soften the equipment operated in the shielded area with respect to its electromagnetic emission.

4. Rate estimates

Our primary physics aim in the experiments discussed here is to reach an excellent statistical accuracy in the measurement of the large cross section for neutral-current neutrino nucleus scattering predicted by Eq. (2). Table 1 summarizes the cross sections for some elements of interest. The rate estimates of Table 2 are made for the neutrino beams at CERN SPS and ACOL, and for the RAL ISIS spallation neutron source and the Gosgen reactor.

Table 2: Neutrino Fluxes, Event Rates and Required Ge Target Masses

Machine	E_ν (GeV)	Flux ($cm^{-2}s^{-1}$)	Rate ($day^{-1}ton^{-1}$)	Ge Target Mass (kg) for rate 1/hour
SPS	<40	$2 \cdot 10^5$	5	5000
ACOL	3	10^7	250	100
ISIS	0.3	10^8	200	120
Reactor	<0.01	$3 \cdot 10^{12}$	60000	0.4

For SPS we have assumed $1.5 \cdot 10^{13}$ protons on target (p.o.t) at 10 s intervals and the neutrino yield $2 \times 10^6\ \nu cm^{-2}/(10^{13}$ p.o.t.) which roughly corresponds to an effective distance of 500 m from the decay channel. The ACOL beam flux is calculated at a distance of 15 m from the decay straight section, using the predicted machine parameters [15]. The ISIS flux is based on a beam current of 100 μA and 15 m distance from the production target. The reactor numbers refer to operation at 3 GW thermal power and 40 m effective distance from the core; such conditions can be met at Gosgen, for example, but larger fluxes may be available elsewhere (at a closer approach to the core).

The required target mass is calculated for Ge calorimeters, asking for an event rate of 1/hour which would allow to collect about 10^4 events in a year. Such an amount of events might enable to study the weak interaction models in detail.

5. Background

The background rate may be 10^4 times higher than the coherent event rate, and must therefore be measured very accurately. This is not a major problem in pulsed beams; continuous sources may require extra care and the use of calorimeter elements of various sizes and shapes. We shall discuss below some of the sources for background, which may turn out to be of prime interest in their own right.

5.1 Quasielastic scattering

As the energy resolution of our projected detector may be rather good, it would be tempting to look for anomalous behaviour in the predicted spectrum of recoil energies. In the upper end of the coherent recoil spectrum it is expected that the nucleus might be left in excited state; this may slightly increase the cross section [11] and, in particular, may deform the detector signal because of the delayed release of the de-excitation γ's from the nucleus. These reactions depend much on the target nucleus and may provide a sensitive test of the models of electroweak interactions [11]. The deformed signals might offer a clear signature and high background rejection in these studies. The normalization with the coherent forward peak could allow another unique determination of the parameters of the electroweak interaction.

5.2 Exotica

Because our method of detection is not based on ionization and charge collection but on the measurement of the heat resulting from the collisional energy loss of the beam particle, the detector could be sensitive to new "exotic" particles escaping all other detectors because of the particle's inability to ionize. Among the candidates there are slow heavy monopoles, slow heavy supersymmetric particles and other candidates for dark matter, for example.

Another interesting kinematic region is the low q^2 part of the recoil spectrum. It has been proposed that a new light neutral gauge boson U could increase the neutrino cross sections at very low q^2 [16]. This gauge boson is necessary if one wants spontaneous breaking of the supersymmetry to generate large masses for spin-0 leptons and quarks at the tree approximation [16].

5.3 Astrophysics and nuclear explosions

The detector with somewhat smaller grains could have interesting applications in astrophysics, for example in the studies of the solar neutrino spectrum or observation of neutrino bursts from supernova events. Also, the detector could be useful in the studies or observations of man-made nuclear explosions.

5.4 Cosmic – ray muons

The sea level rate of 200 $s^{-1}cm^{-2}$ of muons will cause about 1.5 s^{-1} counting rate in 1 kg detector units of about 80 cm^2 area horizontally. The muon will deposit about 10 – 50 MeV in the detectors, which is a clear signature in comparison with the coherent recoils below 1 MeV. The thermal and electric rise time of the detector modules is about 10 μs; this allows the use of a rather narrow window for the vetoing of the coincident events in the detector.

5.5 Residual radioactivity

The α – emitters usually leave narrow lines which may serve for calibration purposes. The β – and γ – emitters are more problematic and must be reduced to a minimum. However, the background is measured to a high accuracy in pulsed beams.

The radioactive background rate must not exceed 10^5/(day·kg) in the coherent recoil energy region below 1 MeV. Basing on experience in low – counting laboratories and on measurements performed on our constructional materials, this background rate seems possible to achieve.

A large calorimetric detector could allow to determine interesting new limits for rare radioactive decays such as the electron decay and neutrinoless double beta decay [5].

5.6 Beam-related particles

The charged particles are easily vetoed because they deposit about 100 MeV in the detector modules in coincidence. Fast neutrons deposit about 30 times less and also cause coincidences. Slower neutrons may be very problematic and their rate may have to be measured separately.

6. Conclusions

We summarize that relatively small (≈ 100 kg) detectors in medium- and low-energy neutrino beams could reach event rates around 1/hour in the coherent forward elastic peak due to neutral currents. This could allow a unique measurement of the parameters $a_0 - a_1/2$ (= 1 in WS model) and $a_0 + a_1/2$ (= $1 - 4\sin^2\theta_W$ in WS model) if the cross section ratios on different nuclei could be accurately determined.

Comparison with other types of events in the same detector and/or other detectors could enable the discrimination among the various gauge field models for electroweak interaction theories.

Any anomalous behaviour in the recoil spectrum could give hints towards exotic phenomena; conversely, the absence of anomalous features could be interpreted to set limits to these phenomena. The selective excitation to nuclear levels could offer another sensitive determination of the weak interaction parameters, if the statistics would allow substantially lower cross sections to be determined.

Acknowledgements: We would like to thank E. Fiorini for helpful discussions on radioactive backgrounds and low-counting experiments. We are grateful to V. Zacek for information on neutrino sources and beams. The helpful advice and encouragement of B. Hyams and K. Winter is gratefully acknowledged.

References

1. D.Z. Freedman, D.N. Schramm and D.L. Tubbs, Ann. Rev. Nucl. Sci. **27**, 167 (1977).

2. T.O. Niinikoski and F. Udo, NP Internal Report 74-6, 1974

3. T.O. Niinikoski, "Cosmic-ray disturbances in thermometry and refrigeration", in **Liquid and Solid Helium**, Halsted (New York 1975) pp. 145-147.

4. S.H. Moseley, J.C. Mather and D. McCammon, J. Appl. Phys. **56**, 1257 (1984)
D. McCammon, S.H. Moseley, J.C. Mather and R.F. Mushotzky, J. Appl. Phys. **56**, 1263 (1984).

5. E. Fiorini and T.O. Niinikoski, Nucl. Inst. Meth. **224**, 83 (1984).

6. T.O. Niinikoski et al., Europhys. Lett. **1**, 499 (1986).

7. A. Alessandrello et al., Proc. Pisa Conference on New Particle Detectors, 1986 (to be published)

8. D. McCammon, M. Juda, J. Zhang, R.L. Kelley, S.H. Moseley and A.E. Szymkowiak, Proc. IEEE Nuclear Science Symposium, San Francisco 1985 (IEEE Trans. Nucl. Sci., to appear).

9. D.Z. Freedman, Phys. Rev. **D9**, 1389 (1974).

10. J. Bernabeu, CERN Report TH.2073-CERN, 8 September 1975.

11. G.J. Gounaris and J.D. Vergados, "Neutrino-Nucleus Reactions and the Structure of Neutral Currents", Univ. Ioannina Report (1977).

12. R. Herman and R. Hofstadter, **High-Energy Electron Scattering Tables**, Stanford Univ. Press (Stanford 1960).

13. A. Drukier and L. Stodolsky, Phys. Rev. **D30**, 2295 (1984).

14. T.O. Niinikoski and A. Rijllart, "Data acquisition and analysis of calorimetric signals", in these Proceedings.

15. V. Zacek, Private communication.

16. P. Fayet, Proc. XVII Rencontre de Moriond, Les Arcs 1982, Ed. J. Tran Thanh Van (Editions Frontieres, Gif-sur-Yvette 1982) Vol.1 p. 483.

Data Acquisition and Analysis of Calorimetric Signals

T.O. Niinikoski and **A. Rijllart**

CERN, CH-1211 Geneva 23, Switzerland

Calorimetric signals have been obtained by irradiating implanted Si and melt doped Ge detectors of a few mm^3 size with an alpha source. A room temperature pre-amplifier and a CAMAC based data acquisition system have been used to resolve the time structure of the signals and to filter them for spectrum analysis. The thermal response is analyzed in terms of a model with two heat reservoirs.

1. Introduction

In order to understand the limiting factors in the decrease of the heat capacity with decreasing temperature of low temperature calorimeters, a detailed study of their thermal response to 6 MeV alpha particles was done. The main contribution to the heat capacity of pure crystalline Si and Ge at low temperatures comes from phonons and is proportional to $(T/\theta_D)^3$, where θ_D is the Debye temperature. At very low temperatures the heat capacity may become dominated by impurities, or by contacts and leads which are necessary to measure the temperature of the device.

A description will be given of the detector biasing, signal amplification, data acquisition and analysis techniques used to determine the various heat capacities and conductivities of the detector assembly. Some conclusions are drawn from our new and earlier published data [1] regarding the excess specific heat and the non-thermal behaviour of the heat transport.

2. Detector biasing and signal amplification

The temperature of the detector was determined by measuring the resistance of a thermistor which had been created by the introduction of suitable impurities, either in a local area or throughout the whole detector. The resistance was measured by applying a voltage on a bias resistor in series with the detector, and measuring the voltage across the detector (see Fig. 1).

To amplify the signal from the detector a JFET input operational amplifier was used with a noise voltage of 1.5 $nV/\sqrt{Hz}$ and a gain of 20 dB operating at room temperature. Further amplification increased the total gain to 73 dB and was also used to create a balanced signal, suitable for transmission across a twisted pair cable of 5 m length to a receiver amplifier (see Fig. 2). The bandwidth of the amplifier chain was chosen to be 0.5 MHz, to correspond to the fastest expected rise times of the signals, limited by the parasitic capacitance of the input lines.

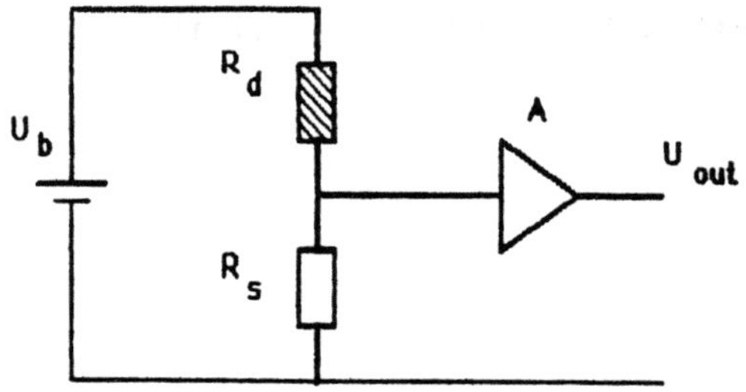

Figure 1. Detector biasing circuit

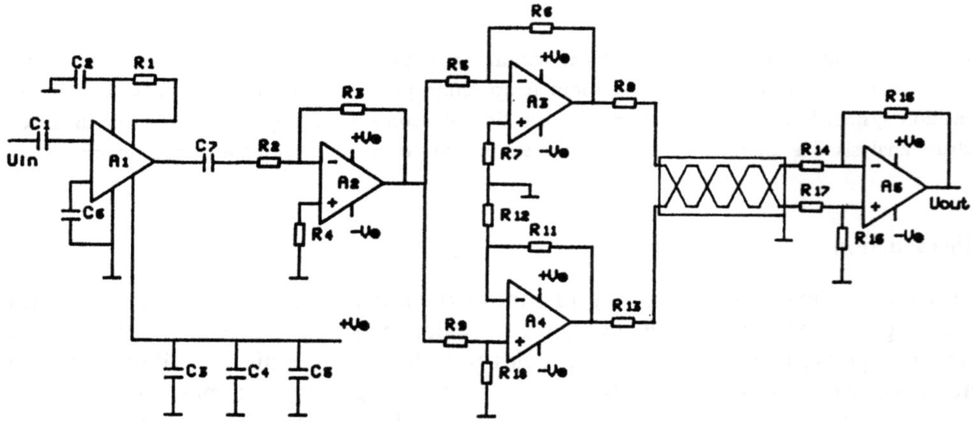

Figure 2. Amplifier circuit

3. Data acquisition and processing

The data acquisition was done using easily available CAMAC equipment for which on-line and off-line software was written. An 8-bit digitizer with a minimum sampling time of 50 ns and a memory depth of 1 kbyte was used to digitize the detector signals generated by the 6 MeV alpha source. The digitized signal was then transferred to a 16-bit Stand-Alone CAMAC processor (MC68000 STAC) [2], which performed on-line analysis, reducing the data typically by a factor of 100. A Super-Caviar microcomputer was used for CAMAC control, histogram storage and display, and off-line analysis. A block diagram of the acquisition system is shown in Fig. 3.

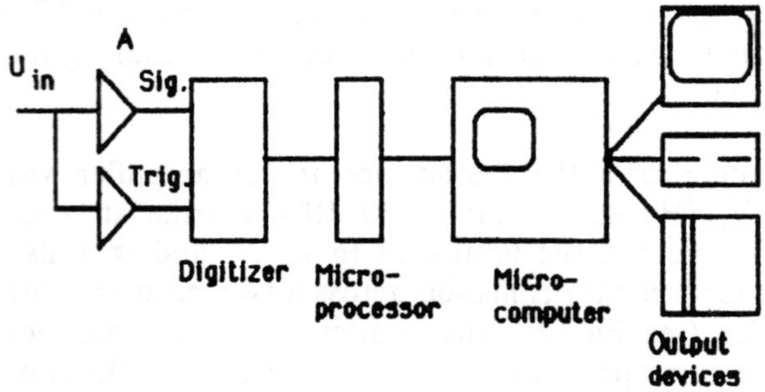

Figure 3. The CAMAC data acquisition system

The on-line data processing consisted of three different algorithms to calculate the pulse height from the digitized signal. The first two algorithms searched for a maximum, one without filtering, the other using low-pass digital filtering. The third, a matched filter algorithm, calculated the maximum using a reference signal, which was created from an averaged signal. This produced three different multi-channel histograms, which provided important information about resolution limiting noise background. The reference signal was also used to perform an rms error calculation, to be able to reject unwanted signals, such as false triggers and double pulses (two pulses very close in time). All accepted signals were averaged to obtain a very high resolution and noise-free signal for off-line time structure analysis at the end of each measurement.

The off-line data analysis included the calculation of rise and decay time constants from averaged signals, and the Fourier analysis of the signals and of the background noise for the purpose of finding the performance and limits of the measurement system.

The measured noise levels, obtained using a random trigger, are compatible with the noise figure of the pre-amplifier.

4. Time structure analysis of calorimetric signals

The decay parts of the pulses obtained by irradiating the detectors with 6 MeV alpha particles show two different time constants in all cases, which can easily be seen in a semi-logarithmic plot of a typical averaged signal, shown in Fig. 4. The vertical axis has an arbitrary logarithmic scale. This behaviour can be described using a two heat reservoir model, shown in Fig. 5, where two heat capacities C_1 and C_2 are in series, interconnected with heat conductivity G_1, and with G_2 connecting C_2 to the copper heat sink. In this model, C_1 represents the heat capacity of the Si or Ge calorimeter with the integrated thermometer, and C_2 the heat capacity of the metallic lead system. G_1 represents the surface heat conductance at the interface between the calorimeter and the leads, and G_2 the heat conductance of the leads.

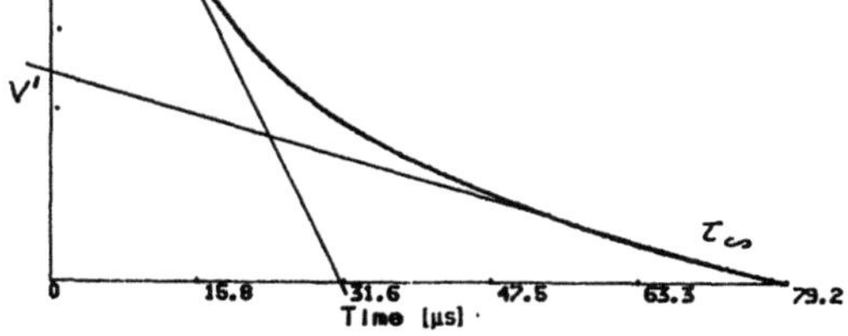

Figure 4. Semi-logarithmic plot of a typical averaged signal

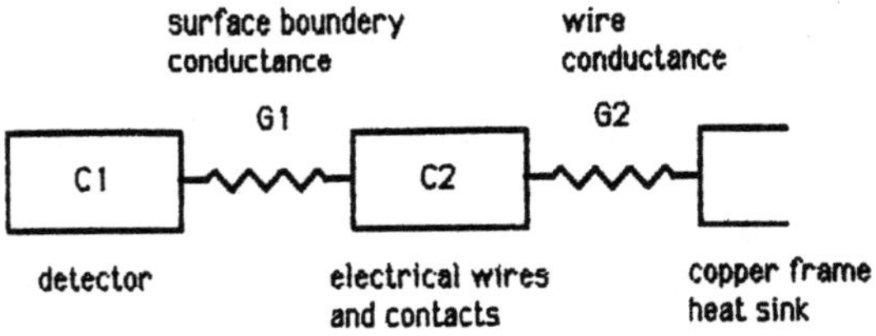

Figure 5. Detector model with two heat reservoirs

This model (in its linearized form) leads to the decay of the temperature pulse with two time constants τ_0 and τ_∞, describing the asymptotic limits of the effective decay time constant at $t=0$ and $t=\infty$, and with two amplitudes V and V′, describing the pulse heights at $t=0$. These amplitudes are obtained by the extrapolation of the initial and final exponential decay curves, as shown in Fig. 4. By denoting

$$U = V'/V \quad \text{and} \quad T = \tau_\infty/\tau_0$$

we can relate the time constants and amplitudes to the parameters of our model:

$$\begin{aligned} C_1 &= \Delta E/\Delta T \propto 1/V \\ G_1 &= C_1/\tau_0 \\ C_2 &= C_1 T^2(1-U)/U(T-1)^2 \\ G_2 &= G_1(T-U)/U(T-1)^2 \end{aligned}$$

When assuming $\tau_\infty \gg \tau_0$, which is often the case above 0.3 K, the last two formulas simplify to

$$\begin{aligned} C_2 &\cong C_1(1-U)/U \\ G_2 &\cong G_1/UT \end{aligned}$$

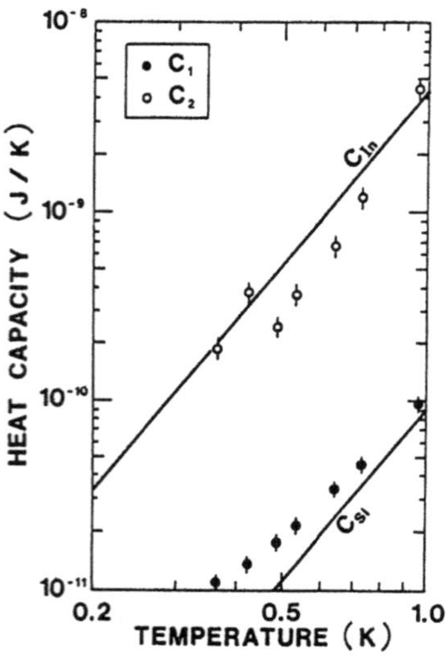

Figure 6. Heat capacities versus temperature

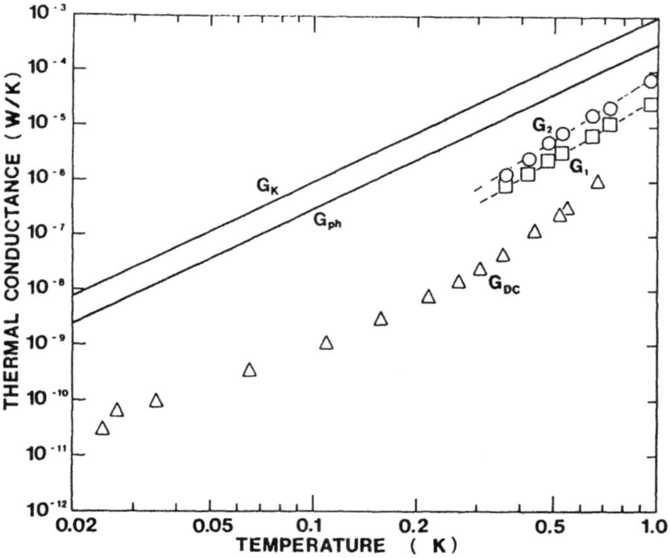

Figure 7. Heat conductances versus temperature

The results for a *Si* :P calorimeter [1] are shown in Figs. 6 and 7. In Fig. 6. we note that C_1 follows closely the C_{Si}, which is the theoretical heat capacity of the detector chip using the Debye specific heat of Si, with deviation growing at lower temperatures. This deviation is expected due to the implanted zones for the thermistor (P impurity) and the ohmic contacts (B impurity). The size of this deviation is about one order of magnitude larger than what one would expect from the data on bulk P−doped Si [3]. This could reflect the influence of the radiation damage due to the implantation which is not entirely annealed away, but could also be due to the much heavier B−doping in the ohmic contacts.

The temperature dependence of C_2 agrees with that of the phonon heat capacity of the superconducting indium leads and solder pads on the aluminium contacts. The magnitude of C_2, however, is only about 20% of the heat capacity of the total amount of indium in the leads and the contacts. The line C_{In} in Fig. 6. corresponds to this 20% of the heat capacity of the total amount of indium.

The above results could be regarded in very good agreement with our thermal model, but unfortunately the dynamic heat conductances G_1 and G_2 are in disagreement with the theoretical ones G_K and G_{Ph} and, in particular, with the measured static heat conductance G_{DC} . These are plotted in Fig. 7. The G_K is the theoretical surface boundary conductance between the Al contact and the Si calorimeter and is obtained using the experimental conductances between metals and dielectric materials. The G_{Ph} is the phonon heat conductance in the In leads and is obtained by using the phonon mean free path of 100μ, which is the thickness of the leads.

5. Discussion

5.1 Heat capacities

We tend to believe that our present results agree with the heat capacities derived from the specific heats of the various constructional materials. Our analysis of the lowest temperature data (not shown here), however, suggests that there are new and so far unknown contributions to the heat capacity. As a possible explanation we would like to point out that the dominant phonon wavelength in Si is $\lambda = 0.3$ mm at 10 mK; this is the thickness of our calorimeter. We could therefore expect that both the phonon specific heat and heat conduction would be dependent on standing wave modes and their couplings among themselves and other heat reservoirs.

Another heat reservoir, which may be significant in small calorimeters, is that of the surface impurities. Oxide layers with H_2O absorbed or adsorbed can present rotational, vibrational and librational degrees of freedom, leading to large heat capacities. Even the clean surface has Rayleigh surface excitations and associated heat capacity; this may also play a role in the heat transmission through interfaces.

Finally, the ^{4}He heat exchange gas usually leaves several monolayers on the surfaces and may be observable at a suitable temperature range in a small calorimeter.

5.2 Heat conduction

The remarkable discrepancy between measured steady-state heat conductivity G_{DC} and the dynamic heat conductivities G_1 and G_2 could be explained in at least two ways:

Firstly, our thermal model may require a correction due to a possible thermal resistance between the thermistor electrons and the bulk phonon system. The electron-phonon heat resistance is size dependent and can be very large in small thermistors. It is then, however, difficult to understand the agreement in the heat capacity measurements, which suggests that the electron-phonon resistance in *Si* :P is small above 0.3 K at least.

Secondly, the electron transport mechanism by hopping and tunneling under the barrier is dependent of the applied electric field. In restricted geometries this effect may become large at low temperatures, because the resistance and voltage drop may concentrate on small critical domains due to the microscopic inhomogeneities in the doping. The result would be an erroneous reading of the calorimeter temperature from calibrations performed using low electric fields.

Other non-thermal phenomena may include ballistic phonons and excitations such as hot electrons or excitons, and slowly relaxing heat from magnetic impurities or vibrational states related with lattice faults and impurities. The superconducting contacts and leads may also trap normal electrons (quasiparticles) which are heated above the phonon temperature; this could offer an explanation towards the difference between conductances in steady-state and in short transients involving non-thermal excitations.

We conclude that although the physics of the thermal detection of ionizing particles seems to be understood with regard to the orders of magnitude of the parameters involved in the temperature range above 0.3 K, there are many phenomena which may merit detailed study already for the sake of their own fundamental interest.

References

1. T.O. Niinikoski, A. Rijllart, A. Alessandrello, E. Fiorini and A. Giuliani : Heat Capacity of a Silicon Calorimeter at Low Temperatures Measured by Alpha Particles, Europhys. Lett. **1**, 499 (1986).

2. A. Rijllart : An MC68000 Stand-Alone CAMAC Microprocessor System, Microprocessing and Microprogramming **12**, 291 (1983).

3. J.R. Marko and J.P. Harrison : Phys. Rev. B **10**, 2448 (1974).

The Use of Rotons in Liquid Helium to Detect Neutrinos

R.E. Lanou, H.J. Maris, and **G.M. Seidel**

Physics Department, Brown University, Providence, RI 02912, USA

Abstract - A new technique for measuring calormetrically the energy spectrum of recoil electrons from the elastic scattering of neutrinos is discussed. The method involves the use of superfluid helium at low temperatures.

1. Introduction

The detection of neutrinos, especially neutrinos having energies in the range of those emanating from the sun, presents experimentalists with a difficult challenge. The importance of understanding the nature of neutrinos has led to a number of ingeneous proposals for their study in recent years. Radiochemical techniques [1], Cerenkov detectors [2], and liquid image chambers [3] are either being proposed or are under construction in order to elucidate the origins of the solar neutrino problem–the lack of agreement between the experimental results of the Davis experiment [4] using $^{37}C\ell$ and the predictions of the standard solar model [5].

Given the paucity of good experimental methods for detecting neutrinos at low energies, calorimetric measurements at low temperature have been proposed as a useful tool for studying the energy spectrum of neutrinos. It is the purpose of this paper to describe a new cryogenic technique, involving only technology and physical principles about which we have a good understanding, that is expected to provide a measurement of the energy spectrum above a few *kev* of recoil electrons resulting from the elastic scattering of neutrinos [6].

2. Cryogenic Detectors

The general approach to detecting neutrinos with cryogenic detectors is to convert the energy of a scattering or absorption event to thermal energy and then to devise some method of measuring the resulting temperature charge. Because the energies involved are so very small, even by low-temperature standards, calorimeters must be developed that have very low heat capacities. One approach to the problem is to make the volume of the material in which the energy is deposited extremely small, as can be achieved with fine superconducting granules. Alternatively, one might try, as Cabrera *et al* [7] have proposed, to work with materials of high Debye temperature.

The problem of measuring the energy of a neutrino event is made doubly difficult by the need for a very large, massive detector to compensate for the extremely weak interaction of neutrinos with matter. As a way around these two contradictory requirements–of small heat capacity and physically large detector–we have developed a scheme whereby the energy released in the neutrino scattering event is transferred from the large detector volume to a small calorimeter with low heat capacity. This transfer is accomplished by the combination of the ballistic propagation of rotons in liquid helium-4 at low temperature and the subsequent evaporation of ^4He atoms from the free surface of the liquid. But before describing details of the technique, it is important to address the problem of background radiation in a practical detector of solar neutrinos.

3. Background Radiation

Without doubt one of the more demanding requirements of a detector for solar neutrinos involves the level of background radiation. Since the number of solar neutrino events is predicted to be extremely small, approximately 1 to $2 per 10^3 kg$-day, background radiation must be kept at a comparable level or else the detector must be able to discriminate between background and neutrino events.

For the purposes of this discussion the background can be divided into two categories, that which propagates from the outside into the detector, and that which originates from the material within the active part of the detector itself. The effects of outside radiation (in a deep mine being principally cosmic ray muons, neutrons from fission events in the surrounding rock, γ's from (n, γ) reactions and radioactive decays in close proximity to the detector) can be shielded against, monitored and distinguished from neutrino events by a variety of techniques. This radiation with its attendant problems is common to all detectors of solar neutrinos, perhaps differing in degree but not in kind. However, what is unique to a particular detector is the radiation arising from the material within it. It is this component of the background which is extremely difficult to distinguish in a calorimetric detector and so must be reduced as much as possible. In this regard we can imagine no substance better suited for neutrino detection than superfluid ^4He. At low temperatures nothing is soluble in liquid helium. For example, a H_2 molecule is calculated [8] to have an energy approximately $100K$ greater in liquid ^4He than in solid hydrogen. At $0.1K$ all foreign material will condense on the container walls. In contrast to solids, liquid helium is a self-cleaning system. Also, the ^4He nucleus has no excited states below $20MeV$ so that neutron reactions are not a problem.

4. Recoil Electrons in ^4He

When a solar neutrino of energy E_ν scatters elastically off a kinetic electron, it imparts an energy to the electron of

$$E_{rec} = \frac{2m_e c^2 E_\nu^2 \cos^2\theta}{E_\nu(1-\cos^2\theta) + 2m_e c^2 E_\nu + m_e c^2} \tag{1}$$

where θ is the angle with respect to the incident neutrino through which the electron is scattered. This energy is transferred to the liquid helium in a few mm by a number of processes; secondary ionizations, the generation of elementary excitations in helium, (phonons and rotons) and the creation of other entities such as metastable He_2 molecules in excited states. Much of the energy that initially goes into producing positive ions and electrons is most likely to end up as rotons when the He^+ ions form He_2^+ complexes and the electrons create bubbles [9]. Some significant fraction of the total energy appears ultimately on de-excitation and recombination as photons,

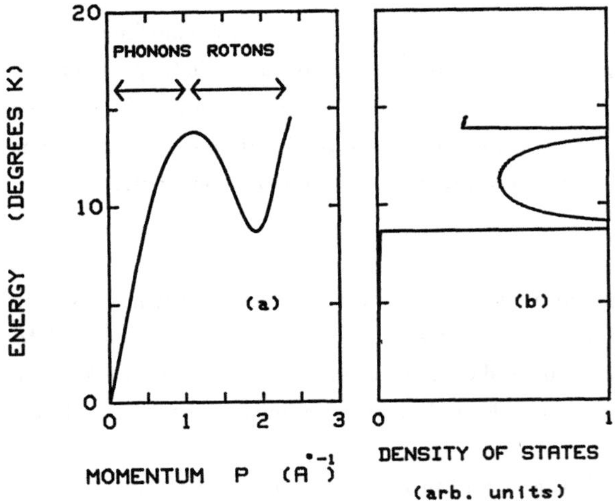

Figure 1. a) Dispersion curve for superfluid helium. b) Density of states of liquid helium in arbitrary units.

principally in the ultraviolet, since liquid helium is known to function as an effective scintillator [10]. Nonetheless, another substantial fraction of the energy must appear finally as elementary excitations of liquid ^4He, the dispersion curve for which is shown in Fig. 1a. Because of the large phase space associated with rotons, as seen in Fig. 1b, almost all this energy will be at large momenta in rotons and high-frequency phonons. No comprehensive measurements appear to have been made on the division of the kinetic energy of an electron in liquid helium, and we assume by way of illustration that the fraction f of that energy appearing as rotons is 0.5. If this fraction turns out to be appreciably smaller, say 0.1, the design of the proposed detector is still valid albeit the measurements become somewhat more difficult.

In liquid helium at low temperatures rotons are essentially stable excitations in that they propagate ballistically large distances through the liquid without scattering or loss of energy [11]. A roton can interact with phonons and other rotons, but at $0.1K$ the thermal equilibrium density of these excitations is sufficiently low that scattering is of no consequence in a volume of reasonable size. At low temperature the time for collision of a roton with thermal phonons is proportional to T^7 and at $0.1K$ is calculated to be $0.05sec$. Such phonon scattering events as do occur are unimportant since there can be little energy or momentum transfer. Also, at low temperature the thermal density of rotons is proportional to $e^{-\Delta/T}$, where $\Delta = 8.65K$ is the energy of the roton minimum (see Fig. 1). For all practical purposes thermal rotons do not exist at $0.1K$, and roton-roton scattering can thus be neglected.

Rotons can scatter as well from ^3He atoms. The normal concentration of ^3He in helium extracted from natural gas wells is 10^{-7}. with a scattering cross section for rotons [13] of approximately $10^{-14}cm^2$ the mean free path is limited to about $0.1cm$. However, the removal of ^3He by standard cryogenic techniques (a superfluid leak) can reduce the concentration of the light isotope by many orders of magnitude.

For an electron recoil energy of $200kev$ the number of rotons produced is, in order of magnitude,

$$N_r = f\frac{E_{rec}}{\Delta} \approx 10^8. \tag{2}$$

On propagating ballistically away from the electron track in all directions, some rotons will hit the free liquid surface directly, while others will first reflect elastically from the confining bottom and side surfaces and eventually reach the free surface. Now the latent heat of a ^4He atom is $7.16K$, a value which is less than the minimum roton energy. Wyatt [14] has shown that a roton on hitting a free surface has a probability of roughly 1/3 of evaporating an atom. Then, in order of magnitude, the number of He atoms N_a evaporated is $N_a \sim 1/3N_r$. For a given recoil energy the precise number will depend upon the geometry of the detector, the location of the event, and two parameters which need to be measured experimentally–the fraction of energy appearing as rotons and the probability for elastic scattering of rotons from surfaces.

Thus, through the creation of rotons, which propagate ballistically and subsequently evaporate He atoms, the detection of neutrinos has been transformed into a measurement of the number of He atoms injected into a vacuum.

5. Detection of He Atoms

The density of ^4He gas in equilibrium with the liquid is extremely small at $0.1K$. At higher temperatures it is known to be given by the relation

$$n_g = 1.5 \times 10^{21} T^{3/2} e^{-7.16/T} \tag{3}$$

which yields at $0.1K$ a value of $10^{-12} cm^{-3}$, not a bad vacuum by terrestrial standards. In thinking about how to measure the number of He atoms evaporated by neutrino event one naturally considers methods involving optical techniques or electronic measurements (such as ionizing the atoms and collecting them), but restrictions imposed by the low temperatures and the need to keep heat inputs very low appear to make practical detectors of this nature difficult. Instead we propose to measure the evaporated atoms calorimetrically by placing above the liquid surface thin plates of a high purity, high Debye temperature, crystalline, dielectric material such as silicon. Attached to the Si wafer will be a sensitive thermometer–a superconducting transition thermometer, a thermistor, or some other device depending on what turns out to be the most suitable. The evaporated atoms will adsorb (physisorb) on the Si surface. Each atom generates in the Si an amount of heat equal to the binding energy ϕ between ^4He and a Si surface. This binding energy is roughly $100K$ and originates in the van der Waals attraction between a He atom and the Si surface [15]. The energy deposited in the Si is then

$$E_{S_i} = \phi N_a \approx \frac{1}{3}\frac{\phi}{\Delta} E_{rec}. \tag{4}$$

Since ϕ/Δ is the order of 10, the energy produced in the Si can be larger than the recoil energy of the electron. The fact that a He atom is bound more strongly to Si than it is to liquid He provides energy amplification.

Silicon has a heat capacity [16] at low temperatures of

$$C = 6.25 T^3 erg/cm^3 K. \tag{5}$$

If the wafer has an area of $200cm^2$ and a thickness of $0.05cm$, then at $0.03K$ a recoil energy of $200kev$ would produced a temperature rise the order of $10^{-3}K$, a value which with current techniques can easily be measured at the 1% level. The energy resolution of the detector is not likely to be limited significantly by the sensitivity of the calorimeter but rather will principally be determined by the dependence of the energy reaching the wafer on the spatial location of the neutrino event.

Since the velocity of rotons is typically $10^4 cm/sec$, the arrival of energy at the wafer will be spread over time the order of $0.01sec$. The time for heat to diffuse in the Si and to achieve thermal equilibrium can be shorter than this, while the time constant for coupling to the heat sink can be adjucted to be longer.

The net result of the transformation is that a neutrino event in a large volume of liquid helium results in the generation of energy in a small silicon wafer. The temperature change in the Si is larger than that which would occur in the helium, were the comparable energy to be thermalized in the bulk liquid, by a factor of at least 10^8.

6. Film Burner

This calorimetric detection scheme carries with it a significant experimental requirement; namely, the surface of the wafer must be kept free of helium. The principal reason He cannot be allowed on the calorimetric wafer is not because the energy amplification (a pleasant but unessential bonus) would be lost, but rather because the heat capacity of a saturated helium film [17] is orders of magnitude larger than that of the Si wafer. Since the silicon is necessarily mounted inside a completely enclosed vessel at $0.1K$, superfluid film flow on the support structure must be "burned off" continuously by heating to about $0.8K$, at the same time maintaining the temperature of the silicon in the range of $.03K$. While the requirements for and construction of film burners are known, inclusion of such a device does add complexity to the detector. The heat necessary to evaporate the film introduces to the liquid helium a large energy input via the recondensing He atoms. The design must be such that the recondensation occurs in a region far removed from the volume of the liquid actively involved in neutrino detection. Also, a carefully designed baffling configuration must ensure that no significant number of ^4He atoms evaporated from the film burner make their way into the high vacuum space where calorimetric detection occurs. This is possible given the greater than 99% probability that when a He atom hits a liquid He surface it will stick [18].

Because the silicon must be heated initially to high temperatures to desorb the gas from the surface while maintaining the helium at very low temperature, the design of the apparatus must allow for thermal isolation between the helium detector and silicon calorimeter. It is therefore reasonable to operate the two subsystems at different temperatures, the helium at $0.1K$ or slightly above, making heat removal easier, and the silicon at approximately $0.03K$ for increased sensitivity.

7. Some Other Considerations

In order to obtain a reasonable number of neutrino events distinguishable from background, liquid helium in a volume of tens of cubic meters needs to be employed. Background γ's from the surrounding material have an attenuation length of 500kev at of approximately 30cm. Dimensions larger than this are required to define a fiducial volume for the separation of signal from background. It is likely that the helium in the detector will be segmented, possibly with thin plastic sheets or other material, each compartment having its own Si wafer or wafers. Details will depend in considerable part on what is learned concerning the reflection of rotons from surfaces.

Within the helium vessel the mass of material (primarily silicon but some structural elements as well) can be kept about 100 times less than that of the helium itself, making the background problem from internal sources tractable.

Liquid helium possesses other properties that may have possible implications for neutrino detection. The lifetime of electrons and He_2^+ ions can be very long in superfluid He such that they can be drifted over long distances and collected, a technique employed in the liquid argon image chamber [19]. However, the sensitivity of present charge-collecting devices does not make such an approach competitive with the proposed calorimetric determination of the recoil energy of electrons produced by solar neutrinos.

The principal source of the luminescence of liquid He at $1K$ produced by 160kev electrons is known to result from the radiative dissociation of the neutral He_2 molecule in the $^3\Sigma_u^+$ state [20]. In fact, roughly 5% of the kinetic energy of a 160kev electron appears as photons between 13 and 18ev. These photons provide an obvious means for studying neutrino scattering events in helium and we are exploring ways in which the proposed calorimetric detector can be made more precise and versatile through the use of information available in the ultraviolet luminescence.

This work was supported in part by Brown University, the U.S. Department of Energy and the National Science Foundation through grants no. DMR 8501858 and DMR 8304224.

8. References

1. W. Hampel, A.I.P. Conf. Proc. **126**, 162 (1985); T. Kirsten, Proc. of IV Moriond Workshop, p. 119 (1986).

2. Sudbury Neutrino Observatory Feasibility Study Report No. SNO-85-3, SNO-86-6; G. Aardsma *et al*, UCI preprint No. 86-47 (also SNO-86-7) "A Heavy Water Detector to Resolve the Solar Neutrino Problem".

3. J. M. Bahcall, M. Baldo-Ceolin, D. B. Cline and C. Rubbia, Phys. Lett. **B178**, 324 (1986).

4. R. Davis, D. S. Harmer, K. C. Hoffman, Phys. Rev. Lett. **20**, 1205 (1968). J. Rowley, B. T. Cleveland, R. Davis in **Solar Neutrinos and Neutrino Astronomy** (Homestake, 1984), ed. by N. Cherry W. Fowler, K. Lande, A.I.P. Cont. Proc. **126**, 1 (1985).

5. J. N. Bahcall *et al*, Rev. Mod. Phys. **54**, 7 67 (1982); J. N. Bahcall, ibid **50** 881 (1978).

6. R. E. Lanou, H. J. Maris and G. M. Seidel, Phys. Rev. Lett. submitted.

7. B. Cabrera, L. M. Krauss and F. Wilczek, Phys. Rev. Lett. **55**, 25 (1985).

8. K. E. Kurten and M. L. Ristig, Phys. Rev. **B31**, 1346 (1985). These authors estimate H_2 has an energy of -$24K$ in liquid He as compared to the vacuum, from which one estimates $100K$ difference from the solid.

9. See A. L. Fetter in **The Physics of Liquid and Solid Helium**, ed. by K. H. Bennemann and J. B. Ketterson (J. Wiley, N.Y. 1976).

10. See. J. B. Birks, **The Theory and Practice of Scintillation Counting** (MacMillan Co., N.Y. 1964).

11. S. Balibar, J. Buechner, B. Castaing, C. Laroche and A. Libchaber, Phys. Rev. **B18**, 3096 (1978).

12. H. J. Maris and R. W. Cline, Phys. Rev. **B23**, 3308 (1981).

13. I. M. Khalatnikov and V. N. Zharkov, Zh. Exsp. Teor. Fiz. **32**, 1108 (1957). Sov. Phys. JETP **5**, 905 (1957).

14. A. F. G. Wyatt, Physics **126B**, 392 (1984).

15. This estimate, by M. J. Cardillo, is an average for different crystallographic orientations.

16. The heat capacity of Si follows a T^3 dependence at least to $0.05K$. W. Knack and M. Meissner in **Proc. of 17th International Conf. on Low Temp. Phys.** (North-Holland, Amsterdam, 1984).

17. Below $0.1K$ the dominant heat capacity of a thick He film arises from ripplons. For a $300A$ film $C = 1.8 \times 10^{-3} T\, erg/cm^2 K$.

18. D. O. Edwards, P. P. Fatouros, G. G. Ihas, P. M. Rozensky, S. Y. Shen and C. P. Tan, Phys. Rev. Lett. **34**, 1153 (1975).

19. H. H. Chen and J. F. Lathrop, Nucl. Instr. and Meth. **150**, 585 (1978); P. J. Doe, H. J. Mahler and H. H. Chen, *ibid* **199**, 639 (1982).

20. M. Stockton, J. K. Keto and W. A. Fitzsimmons, Phys. Rev. Lett. **24**, 654 (1970). C. M. Surko, R. E. Packard, G. J. Dick and F. Reif, Phys. Rev. Lett. **24** 657 (1970).

List of Participants

Barone, A.	University Napels
De Bellefon, A.	College de France, Paris
Bland, R.	San Francisco State University
Booth, N.	University of Oxford
Buschhorn, G.	Max-Planck-Institut für Physik und Astrophysik, München
Camin, D.	University Milano
Carelli, P.	Istituto di Electronica, Roma
Derado, I.	Max-Planck-Institut für Physik und Astrophysik, München
von Feilitzsch, F.	Technische Universität München
Fent, J.	Max-Planck-Institut für Physik und Astrophysik, München
Fiorini, E.	University Milano
Freund, P.	Max-Planck-Institut für Physik und Astrophysik, München
Garcia-Esteve, J.	University Zaragoza
Gonzales-Mestres, L.	CERN
Guiliani, A.	University Milano
Hahn, B.	Universität Bern
Jany, P.	Institut für Kernphysik, Karlsruhe
Klages, H.	Institut für Kernphysik, Karlsruhe
Kotlicki, A.	University Warsaw
Kotthaus, R.	Max-Planck-Institut für Physik und Astrophysik, München
Kraus, H.	Technische Universität München
Liengme, O.	Vandoevres
Morales, A.	University Zaragoza
Neumaier, K.	Walther-Meissner-Insitut für Tieftemperaturforschung, München
Niinikoski, T.	CERN
Pretzl, K.	Max-Planck-Institut für Physik und Astrophysik, München
Pacheco, P.	University Zaragoza
Perret-Gallix, D.	LAPP, Annecy
Petereins, T.	Technische Universität München
Pröbst, F.	Technische Universität München
Rijllart, A.	CERN
Sadoulet, B.	Berkeley
Schmitz, N.	Max-Planck-Institut für Physik und Astrophysik, München
Schouten, J.	Oxford Instruments
Seidel, W.	Technische Universität München
Seidel, G.	Brown University
Seyboth, P.	Max-Planck-Institut für Physik und Astrophysik, München
Singsaas, A.	Max-Planck-Institut für Physik und Astrophysik, München

Stodolsky, L.	Max-Planck-Institut für Physik und Astrophysik, München
Stucki, G.	Universität Bern
Tao, Ch.	College de France, Paris
Vesztergombi, G.	Max-Planck-Institut für Physik und Astrophysik, München
Waysand, G.	University Paris

Index of Contributors

MIX
Papier aus verantwortungsvollen Quellen
Paper from responsible sources
FSC® C105338

If you have any concerns about our products,
you can contact us on
ProductSafety@springernature.com

In case Publisher is established outside the EU,
the EU authorized representative is:
Springer Nature Customer Service Center GmbH
Europaplatz 3, 69115 Heidelberg, Germany

Printed by Libri Plureos GmbH
in Hamburg, Germany